MR. RHEE'S BRILLIANT MATH SERIES

AP CALCULUS

By Brian Rhee

Reflects Changes to the New
2017 AP Exam

45 Topic-Specific Lessons with
Key Summaries

Complete Review of both AP Calculus
AB and BC in this Comprehensive
Test Preparation Book

MR. RHEE'S BRILLIANT MATH SERIES

LEGAL NOTICE

Legal Notice

Copyright © 2016 by Brian Rhee
Published by: Solomon Academy
First Edition
ISBN-13: 978-1541161207
ISBN-10: 1541161203

All rights reserved. This publication or any portion thereof may not be copied, replicated, distributed, or transmitted in any form or by any means whether electronically or mechanically whatsoever. It is illegal to produce derivative works from this publication, in whole or in part, without the prior written permission of the publisher and author.

AP and Advanced Placement Program are registered trademarks of the College board which was not involved in the production of this publication nor endorses this book.

About Author

Brian(Yeon) Rhee obtained a Masters of Arts Degree in Statistics at Columbia University, NY. He served as the Mathematical Statistician at the Bureau of Labor Statistics, DC. He is the Head Academic Director at Solomon Academy due to his devotion to the community coupled with his passion for teaching. His mission is to help students of all confidence level excel in academia to build a strong foundation in character, knowledge, and wisdom. Now, Solomon academy is known as the best academy specialized in Math in Northern Virginia.

Brian Rhee has published five books which are available in www.amazon.com. The titles of his books are AP Calculus, SAT 1 Math, SAT 2 Math level 2, SHSAT/TJHSST Math workbook, and IAAT (Iowa Algebra Aptitude Test). He's currently working on other math books which will be introduced in the near future.

Brian Rhee has twenty years of teaching experience in math. He has been one of the most popular tutors among TJHSST (Thomas Jefferson High School For Science and Technology) students. Currently, he is developing many online math courses with www.masterprep.net for AP Calculus AB and BC, SAT 2 Math level 2 test, and other various math subjects.

Acknowledgements

I wish to acknowledge my deepest appreciation to my wife, Sookyung, who has continuously given me wholehearted support, encouragement, and love. Without you, I could not have completed this book.

Thank you to my sons, Joshua and Jason, who have given me big smiles and inspiration. I love you all.

Thank you to Mr. Kwon from www.Masterprep.net, who has given me opportunities to develop online math courses for various math subjects.

MR. RHEE'S BRILLIANT MATH SERIES   ABOUT THIS BOOK

About This Book

This book is designed to help you master the AP Calculus AB and BC exam. It contains 45 topic-specific lessons with key summaries. Each lesson contains about 5 to 10 practice problems, which are the most up-to-date types of AP Exam test problems.

This book is divided into two parts. The first part consists of lesson 1 through lesson 28 for which are the common topics for AP Calculus AB and BC: limits and continuity, differentiation, applications of derivatives, the definite integral, integration techniques, area between two curves, volume of a solid by revolution, and differential equations.

The second part consists of lesson 29 through lesson 45 for which are the topics for AP Calculus BC only: logarithmic differentiation, L'Hospital's rule, derivatives of parametric and polar equations, volume by cylindrical shells method, integration by parts and partial fractions, improper integral, differential equations including Euler's method and logistic growth model, and sequences and series.

This book is intended for use in AP Calculus AB and BC online courses available in www.masterprep.net. It provides only answer keys for practice problems. All detailed solutions are available to students who register for AP Calculus AB and BC online courses.

MR. RHEE'S BRILLIANT
MATH SERIES

ABOUT AP Exams

ABOUT AP Calculus AB and BC Exams

The AP Calculus AB and BC exams are intended to measure the extent to which a student has mastered the subject matters of the AP Calculus course. Although the AP Calculus courses focus on differential and integral calculus, students need strong foundations in Algebra, Geometry, and Trigonometry.

Each AP exam is 3 hours and 15 minutes long, and its format is as follows:

Section 1   Multiple choice — 45 Questions (1 hour and 45 minutes)
     Part A: 30 questions for 60 minutes (calculator is not permitted)
     Part B: 15 questions for 45 minutes (graphing calculator is required)

Section 2   Free Response — 6 Questions (1 hour and 30 minutes)
     Part A: 2 questions for 30 minutes (graphing calculator is required)
     Part B: 4 questions for 60 minutes (calculator is not permitted)

Contents

Common Topics For AP Calculus AB & BC		9
Lesson 1	The Limit of a Function	11
Lesson 2	Calculating Limits Using the Properties of Limits	17
Lesson 3	Limits at Infinity	23
Lesson 4	Continuity	30
Lesson 5	Average Rate of Change and Instantaneous Rate of Change	36
Lesson 6	Derivatives	43
Lesson 7	Differentiation Rules	48
Lesson 8	Differentiation Rules	54
Lesson 9	The Chain Rule	60
Lesson 10	Implicit Differentiation	66
Lesson 11	Derivatives of Inverse Trig Functions and Higher Derivatives	72
Lesson 12	Indeterminate Forms And L'Hospital's Rule	79
Lesson 13	Related Rates	86
Lesson 14	Linear Approximations And Differentials	92
Lesson 15	Maximum And Minimum Values	99
Lesson 16	The Mean Value Theorem And Rolle's Theorem	106
Lesson 17	Understanding A Curve From The First And Second Derivatives	111
Lesson 18	Optimization Problems	118
Lesson 19	Indefinite Integrals	123
Lesson 20	The Definite Integral	129

MR. RHEE'S BRILLIANT MATH SERIES

TABLE of CONTENTS

Lesson 21	Numerical Approximations Of Integration	135
Lesson 22	The Fundamental Theorem Of Calculus	141
Lesson 23	The U-Substitution Rule	148
Lesson 24	Area Between Curves	155
Lesson 25	Average Value Of A Function and Arc Length	161
Lesson 26	Volumes Of Solids Of Revolution	166
Lesson 27	Volumes Of Solids Of Cross-Sections	174
Lesson 28	Differential Equations	181
AP Calculus BC Topics Only		189
Lesson 29	Logarithmic Differentiation	191
Lesson 30	Indeterminate Products and Indeterminate Powers	196
Lesson 31	Derivative And Arc Length Of Parametric Equations	203
Lesson 32	Volumes By Cylindrical Shells	209
Lesson 33	Integration By Parts	217
Lesson 34	Trigonometric Integrals	223
Lesson 35	Integration By Partial Fractions	229
Lesson 36	Improper Integrals	235
Lesson 37	Differential Equations	241
Lesson 38	Derivative, Arc Length, And Area With Polar Coordinates	249
Lesson 39	Sequences	257
Lesson 40	Convergence And Divergence Of Series, Part I	262
Lesson 41	Convergence And Divergence Of Series, Part II	271
Lesson 42	Strategy For Testing Series	278
Lesson 43	Power Series	284
Lesson 44	Representations Of Functions As Power Series	290
Lesson 45	Taylor And Maclaurin Series	296

COMMON TOPICS

FOR

AB & BC

MR. RHEE'S BRILLIANT MATH SERIES
AB & BC — AP CAL LESSON 1

LESSON 1
The Limit of a Function

Definition of Limit

Let f be a function defined on some open interval that contains the number a, except possibly at a itself. Then we say that the limit of $f(x)$ as x approaches a is L, and we write

$$\lim_{x \to a} f(x) = L$$

if for every number $\varepsilon > 0$ there is a corresponding number $\delta > 0$ such that $|f(x) - L| < \varepsilon$ whenever $0 < |x - a| < \delta$.

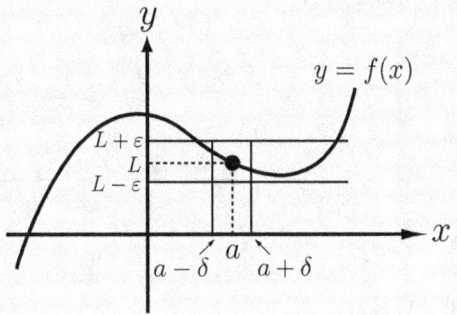

Example 1 Prove the limit by the definition

Prove that $\lim_{x \to 2} 3x - 1 = 5$.

Solution Given $\varepsilon > 0$. If $0 < |x - 2| < \delta$, then

$$|(3x - 1) - 5| < \varepsilon$$
$$|3x - 6| < \varepsilon$$
$$3|x - 2| < \varepsilon$$
$$|x - 2| < \frac{\varepsilon}{3}$$

Since $|x - 2| < \delta$ and $|x - 2| < \frac{\varepsilon}{3}$, choose $\delta = \frac{\varepsilon}{3} > 0$. Thus,

$$|(3x - 1) - 5| < \varepsilon \quad \text{whenever} \quad 0 < |x - 2| < \delta$$

Therefore, by the definition of a limit, $\lim_{x \to 2} 3x - 1 = 5$.

One-Sided Limits

We write
$$\lim_{x \to a^-} f(x) = L$$
and say the **left-hand limit** of $f(x)$ as x approaches a from the left is equal to L as shown in figure 1. Similarly, the **right-hand limit** of $f(x)$ as x approaches a from the right is equal to L as shown in figure 2 and we write
$$\lim_{x \to a^+} f(x) = L$$

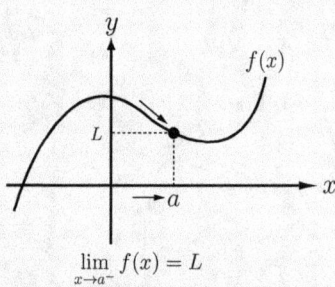

$$\lim_{x \to a^-} f(x) = L$$

Figure 1

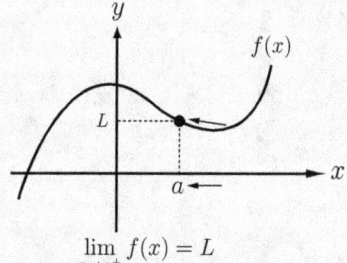

$$\lim_{x \to a^+} f(x) = L$$

Figure 2

When does the Limit Exist?

The limit exists when the left-hand limit and the right-hand limit are the same.

$$\lim_{x \to a} f(x) = L \quad \text{if and only if} \quad \lim_{x \to a^-} f(x) = L \quad \text{and} \quad \lim_{x \to a^+} f(x) = L$$

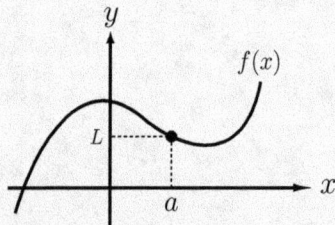

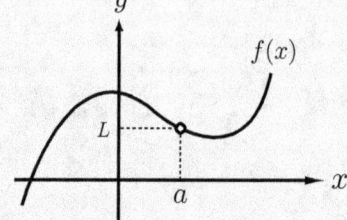

Both graphs above show that the limit exists and $\lim_{x \to a} f(x) = L$.

> **Tip** The limit of a function does not exist for one of three reasons:
>
> 1. The one-sided limits are not the same.
>
> 2. The function does not approach a finite value.
>
> 3. The function oscillates as x approaches a.

Example 2 Having different One-sided limits

Find $\lim_{x \to 3} \dfrac{|x-3|}{x-3}$

Solution The left-hand limit is -1 and the right-hand limit is 1 as shown below.

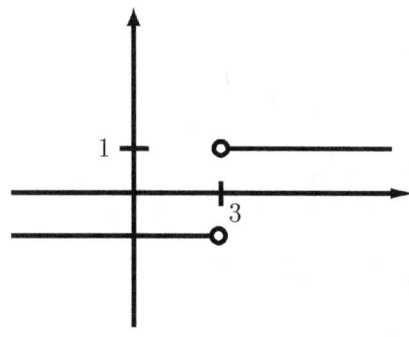

Since the left-hand limit and the right-hand limit are different, the limit does not exist at $x = 3$.

Example 3 Finding the limit of an oscillating function

Find $\lim_{x \to 0} \sin \dfrac{\pi}{x}$.

Solution The value of $\sin \dfrac{\pi}{x}$ oscillates between 1 and -1 infinite often as x approaches 0 as shown below.

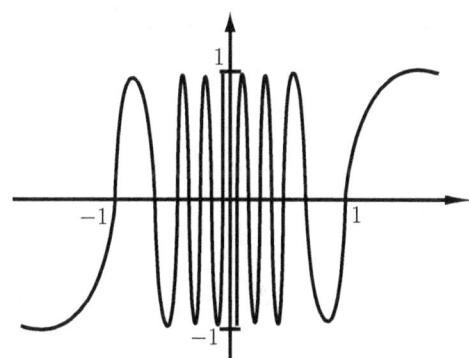

Therefore, $\lim_{x \to 0} \sin \dfrac{\pi}{x}$ does not exist.

MR. RHEE'S BRILLIANT MATH SERIES   AB & BC   AP CAL LESSON 1

EXERCISES

For the function f whose graph is given below, answer questions 1 – 6.

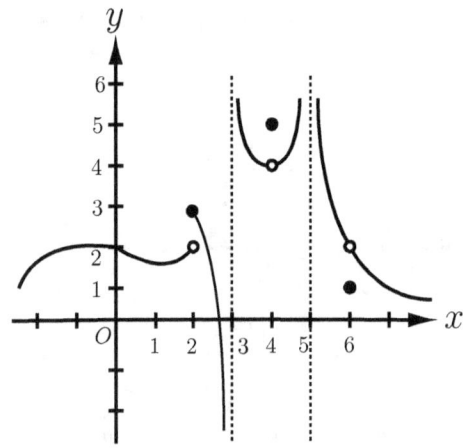

1. Find the value of the following quantity if it exists. If it doesn't exist, explain why.

    (a) $\lim_{x \to 0^-} f(x)$

    (b) $\lim_{x \to 0^+} f(x)$

    (c) $\lim_{x \to 0} f(x)$

    (d) $f(0)$

2. Find the value of the following quantity if it exists. If it doesn't exist, explain why.

    (a) $\lim_{x \to 2^-} f(x)$

    (b) $\lim_{x \to 2^+} f(x)$

    (c) $\lim_{x \to 2} f(x)$

    (d) $f(2)$

3. Find the value of the following quantity if it exists. If it doesn't exist, explain why.

    (a) $\lim_{x \to 3^-} f(x)$

    (b) $\lim_{x \to 3^+} f(x)$

    (c) $\lim_{x \to 3} f(x)$

    (d) $f(3)$

4. Find the value of the following quantity if it exists. If it doesn't exist, explain why.

   (a) $\lim\limits_{x \to 4^-} f(x)$

   (b) $\lim\limits_{x \to 4^+} f(x)$

   (c) $\lim\limits_{x \to 4} f(x)$

   (d) $f(4)$

5. Find the value of the following quantity if it exists. If it doesn't exist, explain why.

   (a) $\lim\limits_{x \to 5^-} f(x)$

   (b) $\lim\limits_{x \to 5^+} f(x)$

   (c) $\lim\limits_{x \to 5} f(x)$

   (d) $f(5)$

6. Find the value of the following quantity if it exists. If it doesn't exist, explain why.

   (a) $\lim\limits_{x \to 6^-} f(x)$

   (b) $\lim\limits_{x \to 6^+} f(x)$

   (c) $\lim\limits_{x \to 6} f(x)$

   (d) $f(6)$

MR. RHEE'S BRILLIANT MATH SERIES — AB & BC — AP CAL LESSON 1

Answers

1. (a) 2   (b) 2   (c) 2   (d) 2
2. (a) 2   (b) 3   (c) does not exist (d) 3
3. (a) $-\infty$ (b) ∞ (c) does not exist (d) does not exist
4. (a) 4   (b) 4   (c) 4   (d) 5
5. (a) ∞ (b) ∞ (c) does not exist (d) does not exist
6. (a) 2   (b) 2   (c) 2   (d) 1

MR. RHEE'S BRILLIANT MATH SERIES AB & BC AP CAL LESSON 2

LESSON 2

Calculating Limits Using the Properties of Limits

Properties of Limits

Suppose that c is a constant and the limits

$$\lim_{x \to a} f(x) \quad \text{and} \quad \lim_{x \to a} g(x)$$

exist. Then use the following properties of limits to calculate limits.

1. $\lim_{x \to a} [f(x) \pm g(x)] = \lim_{x \to a} f(x) \pm \lim_{x \to a} g(x)$

2. $\lim_{x \to a} cf(x) = c \lim_{x \to a} f(x)$

3. $\lim_{x \to a} [f(x) \cdot g(x)] = \lim_{x \to a} f(x) \cdot \lim_{x \to a} g(x)$

4. $\lim_{x \to a} \dfrac{f(x)}{g(x)} = \dfrac{\lim_{x \to a} f(x)}{\lim_{x \to a} g(x)}$

5. $\lim_{x \to a} [f(x)]^n = \left[\lim_{x \to a} f(x) \right]^n$

6. $\lim_{x \to a} c = c$

7. $\lim_{x \to a} x^n = a^n$

8. $\lim_{x \to a} \sqrt[n]{x} = \sqrt[n]{a}$

9. $\lim_{x \to a} \sqrt[n]{f(x)} = \sqrt[n]{\lim_{x \to a} f(x)}$

[Tip]

1. $\lim_{x \to a} = L$ if and only if $\lim_{x \to a^-} = L$ and $\lim_{x \to a^+} = L$

2. If f is a polynomial or rational function and a is in the domain of f, then

$$\lim_{x \to a} f(x) = f(a)$$

MR. RHEE'S BRILLIANT MATH SERIES
AB & BC — AP CAL LESSON 2

Example 1 Finding the limit of a polynomial function

Find $\lim_{x \to 2}(x^2 - 2x - 1)$

Solution

$$\lim_{x \to 2}(x^2 - 2x - 1) = \lim_{x \to 2} x^2 - \lim_{x \to 2} 2x - \lim_{x \to 2} 1$$
$$= \lim_{x \to 2} x^2 - 2\lim_{x \to 2} x - \lim_{x \to 2} 1$$
$$= (2)^2 - 2(2) - 1$$
$$= -1$$

Limit of a Rational Function

In order to find the limit of a rational function $f(x) = \frac{p(x)}{q(x)}$ at $x = a$, plug in $x = a$ to both the numerator and denominator. That is,

$$\lim_{x \to a} \frac{p(x)}{q(x)} = \frac{p(a)}{q(a)}$$

The value of $\frac{p(a)}{q(a)}$ determines the limit of f. Consider the following four cases.

- **Case 1:** If $\frac{p(a)}{q(a)} = \frac{\text{constant}}{\text{constant}}$, the limit is $\frac{\text{constant}}{\text{constant}}$. For instance, let $f(x) = \frac{x-1}{x+1}$.

  $$\lim_{x \to 2} \frac{x-1}{x+1} = \frac{2-1}{2+1} = \frac{1}{3} \quad \Longrightarrow \quad \text{Limit of } f \text{ is } \frac{1}{3}$$

- **Case 2:** If $\frac{p(a)}{q(a)} = \frac{0}{\text{constant}}$, the limit is 0. For instance, let $f(x) = \frac{x-3}{x^2-1}$.

  $$\lim_{x \to 3} \frac{x-3}{x^2-1} = \frac{3-3}{3^2-1} = 0 \quad \Longrightarrow \quad \text{Limit of } f \text{ is } 0.$$

- **Case 3:** If $\frac{p(a)}{q(a)} = \frac{\text{constant}}{0}$, the limit does not exist. For instance, let $f(x) = \frac{1}{(x+1)^2}$.

  $$\lim_{x \to -1} \frac{1}{(x+1)^2} = \frac{1}{(-1+1)^2} = \frac{1}{0} \quad \Longrightarrow \quad \text{Limit of } f \text{ does not exist.}$$

- **Case 4:** If $\frac{p(a)}{q(a)} = \frac{0}{0}$, do the following three extra steps.

  - Step 1: Factor both the numerator and denominator.
  - Step 2: Cancel out a common factor.

– Step 3:   Plug-in $x = a$.

For instance, let $f(x) = \frac{x^2-1}{x-1}$. Since $\lim\limits_{x \to 1} \frac{x^2-1}{x-1} = \frac{0}{0}$, do the following steps.

$$\lim_{x \to 1} \frac{x^2-1}{x-1} = \lim_{x \to 1} \frac{(x+1)(x-1)}{x-1} \qquad \text{Factor the numerator}$$

$$= \lim_{x \to 1} \frac{(x+1)\cancel{(x-1)}}{\cancel{x-1}} \qquad \text{Cancel out } x-1$$

$$= \lim_{x \to 1} (x+1) \qquad \text{Plug-in } x = 1$$

$$= 2$$

Tip $\frac{0}{0}$ or $\frac{\infty}{\infty}$ are called an **indeterminate form**. The limit of the indeterminate form may or may not exist. The limit of the indeterminate form can be easily found using the **L'Hospital's Rule**.

Example 2   Finding the limit of a rational function

Find $\lim\limits_{x \to 0} \frac{\sqrt{x+2} - \sqrt{2}}{x}$

Solution

$$\lim_{x \to 0} \frac{\sqrt{x+2} - \sqrt{2}}{x} = \lim_{x \to 0} \frac{\sqrt{x+2} - \sqrt{2}}{x} \cdot \frac{\sqrt{x+2} + \sqrt{2}}{\sqrt{x+2} + \sqrt{2}}$$

$$= \lim_{x \to 0} \frac{x+2-2}{x(\sqrt{x+2} + \sqrt{2})}$$

$$= \lim_{x \to 0} \frac{x}{x(\sqrt{x+2} + \sqrt{2})}$$

$$= \lim_{x \to 0} \frac{1}{(\sqrt{x+2} + \sqrt{2})}$$

$$= \frac{1}{\sqrt{0+2} + \sqrt{2}}$$

$$= \frac{1}{2\sqrt{2}}$$

MR. RHEE'S BRILLIANT MATH SERIES   AB & BC   AP CAL LESSON 2

Special Limits

The limits of the three functions shown below involve the indeterminate form of $\frac{0}{0}$. Finding the limit of these functions are extremely hard unless you use **L'Hospital's Rule**. Memorize these limits until you learn it.

$$\lim_{x \to 0} \frac{\sin ax}{bx} = \frac{a}{b}, \qquad \lim_{x \to 0} \frac{1 - \cos ax}{bx} = 0 \qquad \lim_{x \to 0} \frac{\tan ax}{bx} = \frac{a}{b}$$

Tip — When you evaluate these limits using a calculator, Make sure to change the Mode to radian. Substitute $x = 0.001$ into the numerator and the denominator to evaluate these limits.

EXERCISES

The graphs of f and g are given below. Use them to answer questions $1-5$.

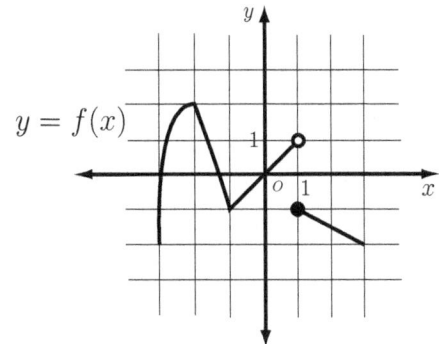
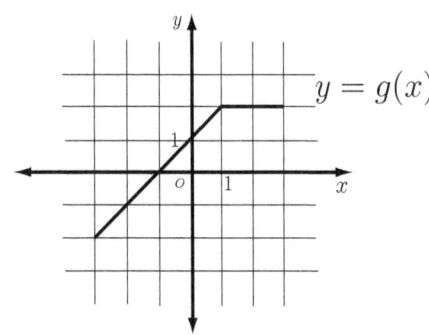

1. Find $\lim\limits_{x \to -2}[f(x) + g(x)]$.

2. Find $\lim\limits_{x \to -1}[f(x)g(x)]$.

3. Find $\lim\limits_{x \to 0} \dfrac{f(x)}{g(x)}$.

4. Find $\lim\limits_{x \to 1} 2f(x)$.

5. Find $\lim\limits_{x \to 2} x^2 g(x)$.

6. Find $\lim\limits_{x \to 2} \sqrt{x+2}$.

7. Find $\lim\limits_{x \to 0} \dfrac{\sin 3x}{2x}$.

8. Find $\lim\limits_{x \to 0} \dfrac{\tan 2x}{x}$.

MR. RHEE'S BRILLIANT MATH SERIES — AB & BC — AP CAL LESSON 2

9. Find $\lim\limits_{x \to -1} \dfrac{\frac{1}{x}+1}{x+1}$.

10. Find $\lim\limits_{h \to 0} \dfrac{(x+h)^2 - x^2}{h}$.

Answers

| 1. 1 | 2. 0 | 3. 0 | 4. does not exist | 5. 8 |
| 6. 2 | 7. $\dfrac{3}{2}$ | 8. 2 | 9. -1 | 10. $2x$ |

LESSON 3

Limits at Infinity

Limits at Infinity

Let f be a function defined on some interval (a, ∞). Then

$$\lim_{x \to \infty} f(x) = L$$

means that the value of $f(x)$ is close to L as x becomes increasing large as shown in Figure 1. Similarly, Let f be a function defined on some interval $(-\infty, a)$. Then

$$\lim_{x \to -\infty} f(x) = L$$

means that the value of $f(x)$ is close to L as x becomes increasing large negative as shown in Figure 2. The line $y = L$ is called a **horizontal asymptote**.

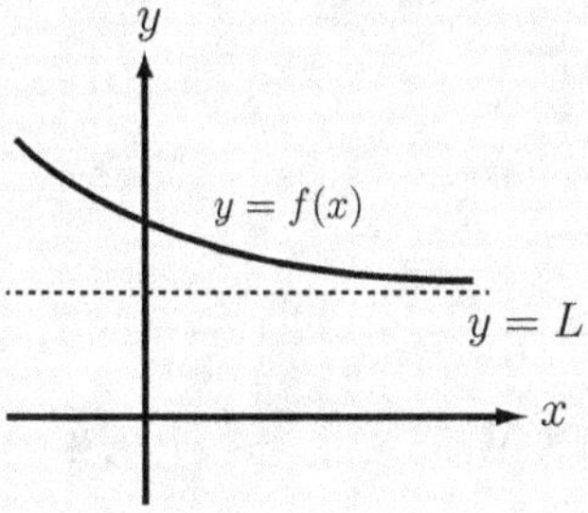

Figure 1 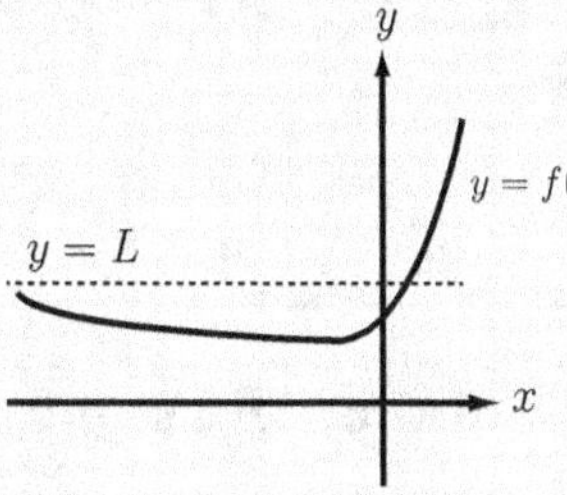   Figure 2

Example 1 Finding the horizontal asymptotes

Find $\lim\limits_{x \to \infty} \tan^{-1} x$ and $\lim\limits_{x \to -\infty} \tan^{-1} x$

Solution

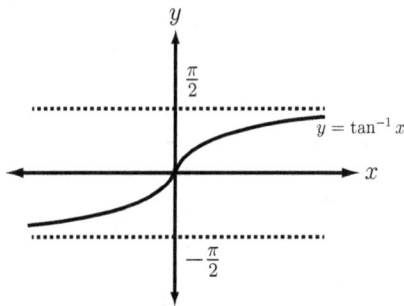

The value of y is close to the line $y = \frac{\pi}{2}$ as x approaches ∞. Whereas, the value of y is close to the line $y = -\frac{\pi}{2}$ as x approaches $-\infty$. Therefore, $\lim_{x \to \infty} \tan^{-1} x = \frac{\pi}{2}$ and $\lim_{x \to -\infty} \tan^{-1} x = -\frac{\pi}{2}$.

$\lim_{x \to \infty} \dfrac{p(x)}{q(x)}$: **Finding Horizontal Asymptote of a Rational Function**

For the rational function

$$f(x) = \frac{p(x)}{q(x)} = \frac{ax^m + \cdots}{bx^n + \cdots}$$

where m is the degree of the numerator and n is the degree of the denominator, a horizontal asymptote can be determined by the following three cases.

- **Case 1:** If $n < m$, there is no horizontal asymptote. In other words, the value of $f(x)$ approaches ∞ or $-\infty$ as x approaches ∞.

- **Case 2:** If $n = m$, f has a horizontal asymptote of $y = \dfrac{a}{b}$, where a and b are the leading coefficients of the numerator and denominator.

- **Case 3:** If $n > m$, f has a horizontal asymptote of $y = 0$.

For instance, for the rational function $f(x) = \frac{1}{x} = \frac{1 \cdot x^0}{x^1}$ whose graph is shown below,

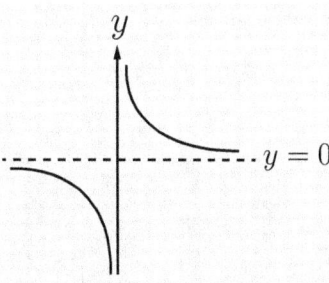

the degree of numerator is 0 and the degree of the denominator is 1. Thus, the rational function has the horizontal asymptote of $y = 0$, which can be denoted by $\lim\limits_{x \to \infty} \dfrac{1}{x} = 0$ and $\lim\limits_{x \to -\infty} \dfrac{1}{x} = 0$.

Example 2 Finding horizontal asymptotes

Find the horizontal asymptotes of each rational function.

(a) $\lim\limits_{x \to \infty} \dfrac{1 - 3x}{2x^2 + 3x}$

(b) $\lim\limits_{x \to \infty} \dfrac{-x^3 + 1}{x^2 - 2x + 2}$

(c) $\lim\limits_{x \to \infty} \dfrac{2x^2 - 2x + 3}{2 - 4x + 3x^2}$

Solution

(a) The numerator is a first degree polynomial ($m = 1$) and the denominator is a second degree polynomial ($n = 2$). Since $n > m$, the horizontal asymptote of f is $y = 0$. Thus, $\lim\limits_{x \to \infty} \dfrac{1 - 3x}{2x^2 + 3x} = 0$.

(b) The numerator is a third degree polynomial ($m = 3$) and the denominator is a second degree polynomial ($n = 2$). Since $n < m$, the value of $f(x)$ approaches $-\infty$ as x approaches ∞. Thus, $\lim\limits_{x \to \infty} \dfrac{-x^3 + 1}{x^2 - 2x + 2} = -\infty$

(c) $f(x) = \dfrac{2x^2 - 2x + 3}{2 - 4x + 3x^2} = \dfrac{2x^2 - 2x + 3}{3x^2 - 4x + 2}$. Both the numerator and the denominator are 2nd degree polynomials. Thus, the horizontal asymptote of f is the ratio of the leading coefficients, or $y = \frac{2}{3}$. Thus, $\lim\limits_{x \to \infty} \dfrac{2x^2 - 2x + 3}{2 - 4x + 3x^2} = \dfrac{2}{3}$.

The definition of Euler's Number, e

$$\lim_{x \to \infty} \left(1 + \frac{1}{x}\right)^x = e \qquad \lim_{x \to \infty} \left(1 + \frac{1}{x}\right)^{cx} = e^c$$

Example 3 Evaluating the limit at infinity

Find the limit of each function.

(a) $\lim\limits_{x \to \infty} \cos x$

(b) $\lim\limits_{x \to -\infty} e^x$

Solution

(a) The value of $\cos x$ oscillate between 1 and -1 infinitely often as x approaches ∞. Thus, $\lim\limits_{x \to \infty} \cos x$ does not exist.

(b) The value of e^x is getting closer to 0 as x approaches $-\infty$. Thus, $\lim\limits_{x \to -\infty} e^x = 0$

MR. RHEE'S BRILLIANT MATH SERIES   AB & BC   AP CAL LESSON 3

EXERCISES

1. Find $\lim_{x \to \infty} \sqrt{x}$.

2. Find $\lim_{x \to \infty} (x^2 - x)$.

3. Find $\lim_{x \to \infty} \sin x$.

4. Find $\lim_{x \to \infty} \dfrac{2x^2 + 1}{1 - x}$.

5. Find $\lim_{x \to \infty} \dfrac{x + 2}{x^3 + 3x^2 + 4}$.

6. Find $\lim_{x \to \infty} \dfrac{7x^2 + 3x + 2}{5x^2 - 2x + 4}$.

MR. RHEE'S BRILLIANT MATH SERIES   AB & BC   AP CAL LESSON 3

7. Find $\displaystyle\lim_{x \to \infty} e^{-x^2}$.

8. Find $\displaystyle\lim_{x \to \infty} \left(1 + \frac{1}{x}\right)^{3x}$.

9. Find $\displaystyle\lim_{x \to \infty} \frac{\sqrt{2x^2 + x}}{x}$.

10. Find $\displaystyle\lim_{x \to -\infty} \frac{\sqrt{2x^2 + x}}{x}$.

MR. RHEE'S BRILLIANT MATH SERIES   AB & BC   AP CAL LESSON 3

Answers

1. ∞ 2. ∞ 3. does not exist   4. $-\infty$ 5. 0
6. $\dfrac{7}{5}$ 7. 0   8. e^3 9. $\sqrt{2}$ 10. $-\sqrt{2}$

LESSON 4

Continuity

A function can be either continuous or discontinuous. The easiest way to test of continuity of a function is to see whether the graph of a function can be traced with a pen without lifting the pen from the paper. In general, mathematical proof of continuity of a function can be done using the concepts of limits.

Definition of Continuity

A function f is continuous at $x = a$ if

$$\lim_{x \to a} f(x) = f(a)$$

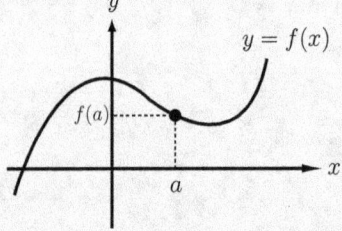

Figure 1

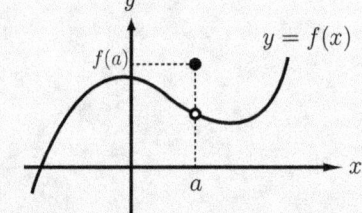

Figure 2

In Figure 1, $\lim_{x \to a} f(x)$ exists and $f(a)$ is defined. Since $\lim_{x \to a} f(x) = f(a)$, the function f is continuous at $x = a$. Whereas, the function f in Figure 2 is discontinuous at $x = a$ because $\lim_{x \to a} f(x) \neq f(a)$.

Tip Testing continuity of a function at $x = a$ requires three steps.

1. Check whether $\lim_{x \to a} f(x)$ exists.

2. Check whether $f(a)$ is defined.

3. Check whether $\lim_{x \to a} f(x) = f(a)$.

Example 1 Finding discontinuity

If $f(x) = \dfrac{x^2 - 2x - 3}{x - 3}$, where is f discontinuous?

Solution

$$\frac{x^2 - 2x - 3}{x - 3} = \frac{(x+1)(x-3)}{x-3}$$
$$= x + 1$$

f has a common factor $x - 3$ from the numerator and the denominator. Cancelling the common factor creates a hole on the graph of f at $x = 3$ as shown below.

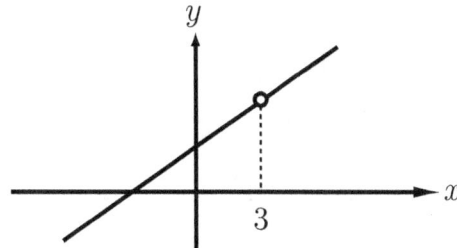

Thus, f is discontinuous at $x = 3$.

Types of Discontinuities

A function f is discontinuous at $x = a$ if f has either a hole, a jump, or vertical asymptote at $x = a$ on its graph.

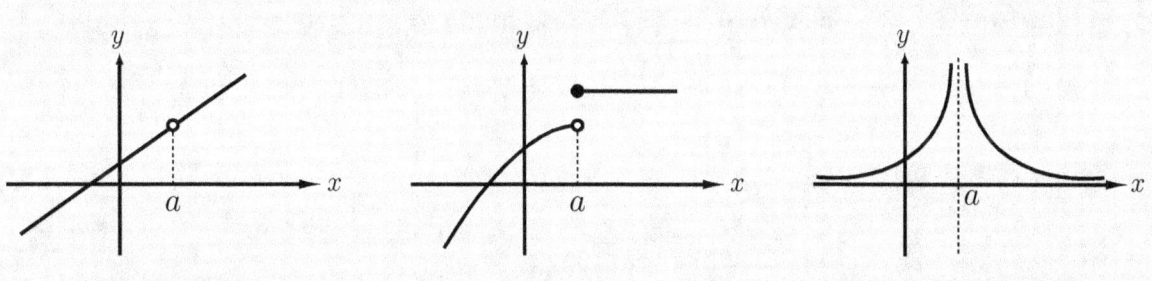

Figure 1: Hole   Figure 2: Jump   Figure 3: Vertical asymptote

First discontinuity illustrated in Figure 1 is called **removable discontinuity** because we can remove the discontinuity by redefining f. In other words, a removable discontinuity is a point at which the graph is not connected but can be connected by filling in a single point. Second discontinuity illustrated in Figure 2 is called **jump discontinuity** because the function jumps from one value to another. Third discontinuity illustrated in Figure 3 is called **infinite discontinuity** because the function has a vertical asymptote. The jump and infinite discontinuities are called **non-removable discontinuity** because the graph can not be connected by filling in a single point.

MR. RHEE'S BRILLIANT MATH SERIES — AB & BC — AP CAL LESSON 4

An important characteristic of continuous functions is shown in the following theorem.

The Intermediate Value Theorem

Suppose that f is continuous on the closed interval $[a, b]$ and let N be any number between $f(a)$ and $f(b)$. Then there exists a number c in (a, b) such that $f(c) = N$.

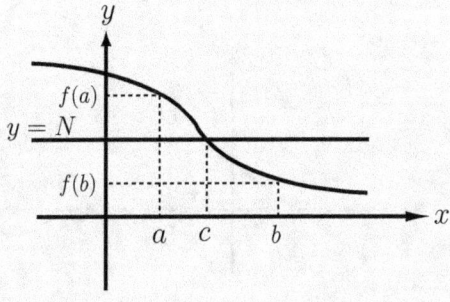

Tip
1. Intermediate value theorem is true for any continuous functions.
2. The use of the Intermediate value theorem is to find a zero of a continuous function: If $f(a)$ and $f(b)$ are of opposite signs, there must be at least one zero between a and b.

Example 2   Finding a zero using the Intermediate value theorem

Show that $f(x) = 2x^3 + 7x^2 - 3x - 18$ has a zero between 1 and 2.

Solution
Evaluate $f(1)$ and $f(2)$.

$$f(1) = 2(1)^3 + 7(1)^2 - 3(1) - 18 = -12$$
$$f(2) = 2(2)^3 + 7(2)^2 - 3(2) - 18 = 20$$

Since $f(1)$ and $f(2)$ are of opposite signs, there must be a zero between 1 and 2.

MR. RHEE'S BRILLIANT MATH SERIES AB & BC AP CAL LESSON 4

EXERCISES

For questions 1-4, sketch the graph of the function and describe the discontinuity.

1. $f(x) = \dfrac{x^2 + 2x + 1}{x + 1}$

2. $f(x) = \dfrac{2x - 5}{|5 - 2x|}$

3. $f(x) = \begin{cases} \sqrt{x - 1}, & x \geq 1 \\ (x - 1)^2 - 1, & x < 1 \end{cases}$

4. $f(x) = \dfrac{x - 2}{x^2 - x - 2}$

MR. RHEE'S BRILLIANT MATH SERIES — AB & BC — AP CAL LESSON 4

5. Show that $f(x) = 2x^3 - 13x^2 + 13x + 10$ has a zero on the interval $[-1, 0]$.

6. Find the value of c that makes f continuous on the interval $[-\infty, \infty]$

$$f(x) = \begin{cases} 2x + 6, & x \leq -1 \\ -x + c, & x > -1 \end{cases}$$

7. Find the value of c that makes f continuous for all x.

$$f(x) = \begin{cases} \dfrac{x^2 - 9}{x - 3}, & x \neq 3 \\ c, & x = 3 \end{cases}$$

8. Find the value of c that makes f continuous on the interval $[-\infty, \infty]$

$$f(x) = \begin{cases} cx + 1, & x \leq 2 \\ cx^2 - 9, & x > 2 \end{cases}$$

MR. RHEE'S BRILLIANT MATH SERIES
AB & BC — AP CAL LESSON 4

Answers

1. Hole(removable discontinuity)
2. Jump discontinuity
3. Jump discontinuity
4. Hole and infinite discontinuity
5. $f(-1) = -18$, $f(0) = 10$
6. $c = 3$
7. $c = 6$
8. $c = 5$

LESSON 5

Average Rate of Change and Instantaneous Rate of Change

Rate of Change

Average rate change measures how much f changes over an interval from $x = a$ to $x = a + h$. Thus, average rate of change is defined as

$$\text{Average rate of change} = \frac{\Delta f}{\Delta x} = \frac{f(a+h) - f(a)}{h}$$

Notice that the average rate of change is the slope of the secant line in Figure 1. Whereas, **instantaneous rate of change** measures how much f changes over a very short interval ($h \approx 0$). Thus, instantaneous rate of change is defined as

$$\text{Instantaneous rate of change} = \lim_{h \to 0} \frac{\Delta f}{\Delta x} = \lim_{h \to 0} \frac{f(a+h) - f(a)}{h}$$

Notice that the instantaneous rate of change is the slope of the tangent line in Figure 2.

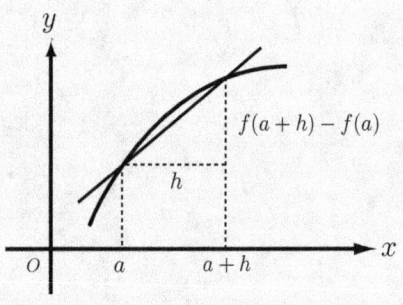
Figure 1

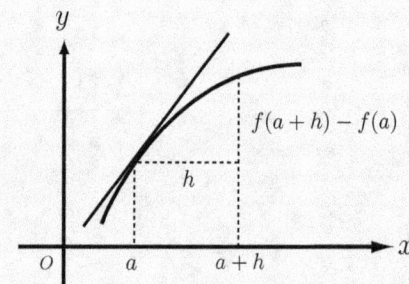
Figure 2

Tip

1. A **secant line** is a straight line that joins two points on a function.

2. A **tangent line** is a straight line that touches the function at a point without crossing over.

3. Instantaneous rate of change can be expressed by one of the two forms shown below.

$$\lim_{h \to 0} \frac{f(a+h) - f(a)}{h} \quad \text{or} \quad \lim_{x \to a} \frac{f(x) - f(a)}{x - a}$$

Tangent Line

The tangent line to the curve $y = f(x)$ at the point $(a, f(a))$ is the line with slope m where

$$m = \lim_{h \to 0} \frac{f(a+h) - f(a)}{h}$$

Thus, the equation of the tangent line at the point $(a, f(a))$ is as follows:

$$y - f(a) = m(x - a), \qquad \text{where} \quad m = \lim_{h \to 0} \frac{f(a+h) - f(a)}{h}$$

Tip Point-slope form of a line through the point (x_0, y_0) with slope m is as follows:

$$y - y_0 = m(x - x_0)$$

Example 1 Writing an equation of the tangent line

Find an equation of the tangent line to $y = x^2$ at the point $(2, 4)$.

Solution Let's find the slope of the tangent line first. Since $f(x) = x^2$ and $a = 2$,

$$\begin{aligned}
m &= \lim_{h \to 0} \frac{f(a+h) - f(a)}{h} \\
&= \lim_{h \to 0} \frac{f(2+h) - f(2)}{h} = \lim_{h \to 0} \frac{(2+h)^2 - 4}{h} \\
&= \lim_{h \to 0} \frac{(4 + 4h + h^2) - 4}{h} = \lim_{h \to 0} \frac{4h + h^2}{h} \\
&= \lim_{h \to 0} \frac{h(4+h)}{h} = \lim_{h \to 0} (4 + h) \\
&= 4
\end{aligned}$$

the slope of the tangent line is 4. Thus, the equation of the tangent to $y = x^2$ at the point $(2, 4)$ is

$$y - 4 = 4(x - 2) \qquad \text{or} \qquad y = 4x - 4$$

MR. RHEE'S BRILLIANT MATH SERIES — AB & BC — AP CAL LESSON 5

Average Velocity and Instantaneous Velocity

Average velocity is change in position over time interval from $t = a$ to $t = a + h$. Let $f(x)$ be the function that describes the position of object at time t. Average velocity is defined as

$$\text{Average velocity} = \frac{\text{displacement}}{\text{time}} = \frac{f(a+h) - f(a)}{h}$$

The average velocity is the slope of the secant line in Figure 3. Whereas, **Velocity**(or **instantaneous velocity**) is change in position over very short time interval($h \approx 0$). Velocity is defined as

$$\text{Velocity} = \lim_{h \to 0} \frac{f(a+h) - f(a)}{h}$$

The velocity is the slope of the tangent line in Figure 4.

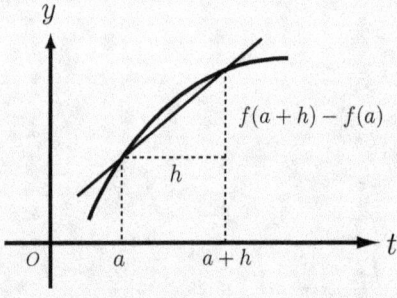

Figure 3

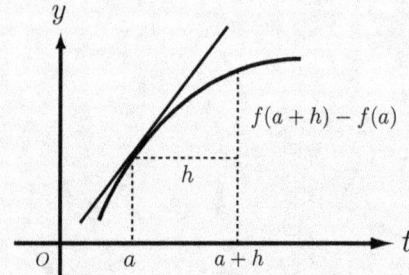

Figure 4

Example 2 Finding velocities

The displacement(in feet) of a particle moving in a straight line is $s(t) = t^2 - 2t + 3$, where t is measured in seconds.

(a) Find the average velocity from $t = 1$ to $t = 3$.

(b) Find the velocity when $t = 4$.

Solution

(a) Since $s(3) = 6$ and $s(1) = 2$,

$$\text{Average velocity} = \frac{s(a+h) - s(a)}{h}$$
$$= \frac{s(3) - s(1)}{2} = \frac{6-2}{2}$$
$$= 2$$

the average velocity over from $t = 1$ to $t = 3$ is 2 ft/s.

(b) In order to find the velocity when $t = 4$, substitute 4 for a.

$$\text{Velocity} = \lim_{h \to 0} \frac{s(a+h) - s(a)}{h} = \lim_{h \to 0} \frac{s(4+h) - s(4)}{h}$$
$$= \lim_{h \to 0} \frac{((4+h)^2 - 2(4+h) + 3) - (4^2 - 2(4) + 3)}{h}$$
$$= \lim_{h \to 0} \frac{6h + h^2}{h} = \lim_{h \to 0} \frac{h(6+h)}{h}$$
$$= \lim_{h \to 0} (6+h) = 6$$

Thus, the velocity when $t = 4$ is 6 ft/s.

MR. RHEE'S BRILLIANT MATH SERIES AB & BC   AP CAL LESSON 5

EXERCISES

The population(in thousands) of Fairfax, Virginia from 2004 to 2015 is given in the table below. Use the table to answer questions 1-5.

Year	2004	2006	2007	2010	2013	2015
Population	235	246	262	298	287	289

1. Find the average rate of growth from 2004 to 2006.

2. Find the average rate of growth from 2007 to 2015.

3. Find the average rate of growth from 2004 to 2015.

4. Estimate the instantaneous rate of growth in 2006.

5. Estimate the instantaneous rate of growth in 2010.

MR. RHEE'S BRILLIANT MATH SERIES — AB & BC — AP CAL LESSON 5

6. Find the slope of the tangent line to $y = \frac{1}{x}$ at the point $(1,1)$.

7. Find an equation of the tangent line to $y = \sqrt{x}$ at the point $(4,2)$.

8. Find an equation of the tangent line to $y = x^2 - 2x$ at the point $(1,-1)$.

For questions 9-10. If a ball is thrown into air with a velocity of 25 m/s, its height (in meters) after t seconds is given by $y = 10t - 2t^2$.

9. When will the ball hit the ground?

10. Find the velocity of the ball after 3 seconds.

MR. RHEE'S BRILLIANT MATH SERIES — AB & BC — AP CAL LESSON 5

Answers

1. 5.5 thousands/yr
2. 3.375 thousands/yr
3. 4.91 thousands/yr
4. 16 thousands/yr
5. -3.67 thousands/yr
6. -1
7. $y - 2 = \dfrac{1}{4}(x - 4)$
8. $y = -1$
9. 5 seconds
10. -2 m/s

MR. RHEE'S BRILLIANT MATH SERIES AB & BC   AP CAL LESSON 6

LESSON 6

Derivatives

Definition of the Derivative Function

The definition of the **derivative of** f is defined as

$$f'(x) = \lim_{h \to 0} \frac{f(x+h) - f(x)}{h}$$

and $f'(x)$ is read as f prime of x. The derivative of f is a function that determines the slope of the tangent lines to a graph of f. Similarly, $f(a)$ is defined as

$$f'(a) = \lim_{h \to 0} \frac{f(a+h) - f(a)}{h}$$

and $f'(a)$ determines the slope of the tangent to the graph of f at $x = a$.

(Tip) The common notations for the derivative are $f'(x)$, $\frac{dy}{dx}$, y', $\frac{d}{dx}f(x)$.

Example 1 Finding the derivative function

Find the derivative function of $f(x) = \sqrt{x}$ and evaluate $f'(4)$.

Solution Let's find $f'(x)$ using the definition of the derivative function.

$$\begin{aligned} f'(x) &= \lim_{h \to 0} \frac{f(x+h) - f(x)}{h} \\ &= \lim_{h \to 0} \frac{\sqrt{x+h} - \sqrt{x}}{h} \\ &= \lim_{h \to 0} \frac{\sqrt{x+h} - \sqrt{x}}{h} \cdot \frac{\sqrt{x+h} + \sqrt{x}}{\sqrt{x+h} + \sqrt{x}} \\ &= \lim_{h \to 0} \frac{x+h-x}{h(\sqrt{x+h} + \sqrt{x})} = \lim_{h \to 0} \frac{h}{h(\sqrt{x+h} + \sqrt{x})} \\ &= \lim_{h \to 0} \frac{1}{\sqrt{x+h} + \sqrt{x}} = \frac{1}{\sqrt{x+0} + \sqrt{x}} \\ &= \frac{1}{2\sqrt{x}} \end{aligned}$$

Since $f'(x) = \frac{1}{2\sqrt{x}}$, $f'(4) = \frac{1}{2\sqrt{4}} = \frac{1}{4}$.

How to Sketch the Graph of the Derivative Function?

Draw many tangent lines to a curve $y = f(x)$ and estimate the slope of each tangent line as shown in Figure 1. For instance, we draw the tangent line to the curve at $x = -3$ and estimate its slope to be about 2. So, $f'(-3) = 2$.

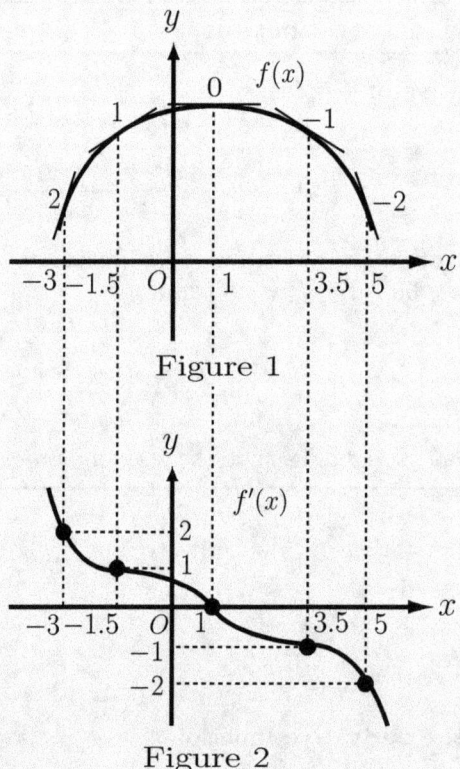

Figure 1

Figure 2

Plot the point $(-3, 2)$ in Figure 2. Repeat this procedure at several points. Since $f'(-1.5) = 1$, $f'(1) = 0$, $f'(3.5) = -1$, and $f'(5) = -2$, plot the points $(-1.5, 1)$, $(1, 0)$, $(3.5, -1)$, and $(5, -2)$ in Figure 2. Finally, connect the points to sketch the graph of $f'(x)$.

Differentiability

A function f is differentiable at $x = a$ if $f'(a)$ exists. It is differentiable on an open interval (a, b) if it is differentiable at every point in the interval. A graph of a differentiable function on the interval (a, b) looks like a continuous smooth curve as shown below.

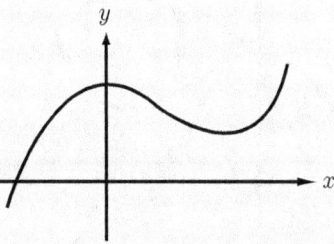

Tip
1. If f is differentiable at $x = a$, then f is continuous at $x = a$.
2. Although f is continuous at $x = a$, it does not mean that f is differentiable at $x = a$.

Non Differentiable Functions

Some functions are not differentiable at $x = a$. f is said to be a non differentiable function at $x = a$ if it has a hole, jump, sharp corner, vertical asymptote, and cusp at $x = a$ as shown below.

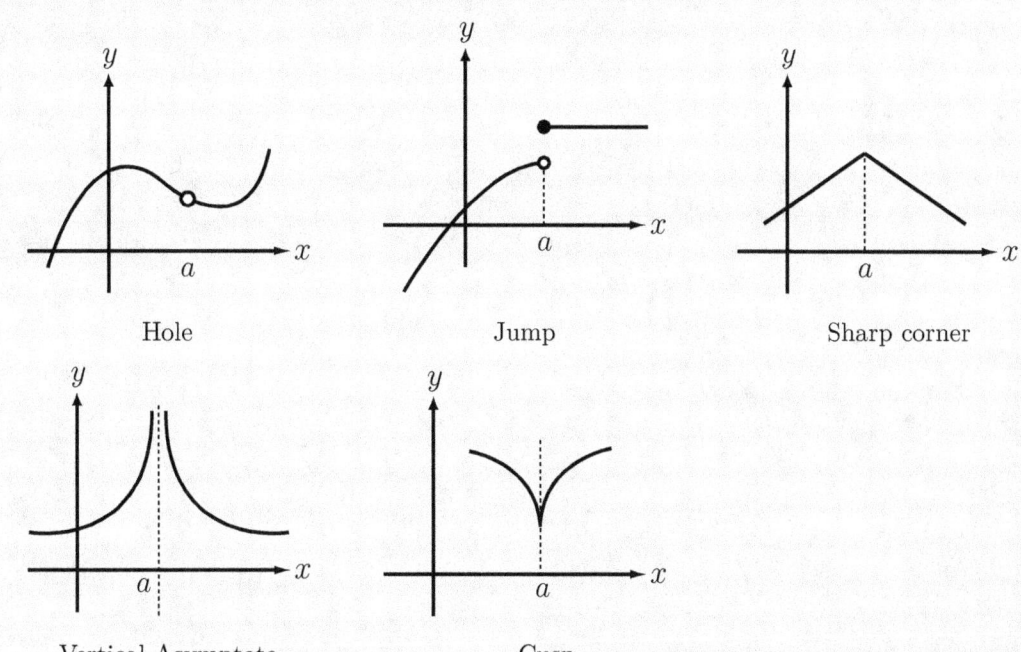

Tip Cusp shown above has a vertical tangent at $x = a$. Since the slope of the vertical tangent is undefined, cusp is not differentiable at $x = a$.

EXERCISES

1. Sketch the graph of the derivative function of $f(x)$ shown below.

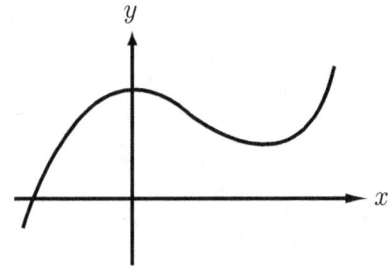

For questions 2-5, find the derivative function using the definition of the derivative and evaluate $f'(3)$.

2. $f(x) = 2x + 3$

3. $f(x) = 4 - x^2$

4. $f(x) = \dfrac{1}{x}$

5. $f(x) = x^3$

MR. RHEE'S BRILLIANT MATH SERIES AB & BC   AP CAL LESSON 6

Answers

1. Graph will be shown in the lecture.
2. $f'(x) = 2$, $\quad f'(3) = 2$
3. $f'(x) = -2x$, $\quad f'(3) = -6$
4. $f'(x) = -\dfrac{1}{x^2}$, $\quad f'(3) = -\dfrac{1}{9}$
5. $f'(x) = 3x^2$, $\quad f'(3) = 27$

LESSON 7

Differentiation Rules

Differentiation means finding the derivative function. In general, finding the derivative function using the definition of the derivative is very tedious and takes long time. In this lesson, you will learn about the differentiation rules without using the definition of the derivative so that you can find the derivative function with ease.

Differentiation Rules

The table below summarizes the differentiation rules.

Differentiation rules	Example
1. $\frac{d}{dx}(c) = 0$	1. $\frac{d}{dx}(2) = 0$
2. $\frac{d}{dx}cf(x) = c \cdot \frac{d}{dx}f(x)$	2. $\frac{d}{dx}2x^2 = 2 \cdot \frac{d}{dx}x^2$
3. $\frac{d}{dx}x^n = nx^{n-1}$	3. $\frac{d}{dx}x^3 = 3x^{3-1} = 3x^2$
4. $\frac{d}{dx}[f(x) \pm g(x)] = \frac{d}{dx}f(x) \pm \frac{d}{dx}g(x)$	4. $\frac{d}{dx}(x^3 + x) = \frac{d}{dx}x^3 + \frac{d}{dx}x = 3x^2 + 1$
5. $\frac{d}{dx}a^x = a^x \cdot \ln a$	5. $\frac{d}{dx}e^x = e^x \cdot \ln e = e^x$
6. $\frac{d}{dx}\log_a x = \frac{1}{x} \cdot \ln a$	6. $\frac{d}{dx}\ln x = \frac{1}{x} \cdot \ln e = \frac{1}{x}$

Tip

1. Note that $\frac{d}{dx}cf(x) \neq \frac{d}{dx}c \cdot \frac{d}{dx}f(x)$. For instance,

$$\frac{d}{dx}2x^2 \neq \frac{d}{dx}2 \cdot \frac{d}{dx}x^2$$

2. The 3rd differentiation rule shown above is called the **Power Rule**. You can apply the power rule to any power function $f(x) = x^n$, where the base is variable and exponent is constant.. The derivative function of the following power functions are worth memorizing.

$$\frac{d}{dx}x = 1, \quad \frac{d}{dx}x^2 = 2x, \quad \frac{d}{dx}x^3 = 3x^2, \quad \frac{d}{dx}\sqrt{x} = \frac{1}{2\sqrt{x}}, \quad \frac{d}{dx}\frac{1}{x} = -\frac{1}{x^2}$$

Example 1 Applying the differentiation rules

Differentiate $f(x) = 2x^3$.

Solution

$$f'(x) = 2(x^3)' = 2(3x^2) = 6x^2$$

Example 2 Applying the differentiation rules

Differentiate $f(x) = (x-1)^2$.

Solution Since $(x-1)^2 = x^2 - 2x + 1$, differentiate each term.

$$\begin{aligned}f'(x) &= (x^2 - 2x + 1)' \\ &= (x^2)' - 2(x)' + (1)' \\ &= 2x - 2(1) + 0 \\ &= 2x - 2\end{aligned}$$

Example 3 Applying the differentiation rules

Differentiate $f(x) = \sqrt{x}$.

Solution Since $\sqrt{x} = x^{\frac{1}{2}}$, apply the power rule.

$$\begin{aligned}f'(x) &= (\sqrt{x})' = (x^{\frac{1}{2}})' \\ &= \frac{1}{2} \cdot x^{\frac{1}{2}-1} = \frac{1}{2} \cdot x^{-\frac{1}{2}} \\ &= \frac{1}{2} \cdot \frac{1}{\sqrt{x}} = \frac{1}{2\sqrt{x}}\end{aligned}$$

MR. RHEE'S BRILLIANT MATH SERIES — AB & BC — AP CAL LESSON 7

Example 4 Finding the tangent line

Find an equation of the tangent line to the curve $f(x) = \frac{1}{x}$ at $(2, \frac{1}{2})$.

Solution Find the derivative function of $f(x) = \frac{1}{x}$ using the power rule.

$$f'(x) = \left(\frac{1}{x}\right)' = (x^{-1})'$$
$$= -1 \cdot x^{-2} = -\frac{1}{x^2}$$

The slope of the tangent line at $x = 2$ is $f'(2) = -\frac{1}{4}$. Since the point of tangency is $(2, \frac{1}{2})$, the equation of the tangent line in point-slope form is

$$y - \frac{1}{2} = -\frac{1}{4}(x - 2)$$

MR. RHEE'S BRILLIANT MATH SERIES   AB & BC   AP CAL LESSON 7

EXERCISES

For questions 1-6, differentiate the following functions.

1. $f(x) = 2x^2 - 3x + 1$

2. $f(x) = (x-1)(x-2)$

3. $f(x) = \dfrac{3}{\sqrt{x}}$

4. $f(x) = \dfrac{1}{x} - \dfrac{2}{x^2} + \dfrac{3}{x^3}$

5. $f(x) = \ln x + ex$

6. $f(x) = \sqrt{x} + \sqrt[3]{x^2} + \sqrt[4]{x^3}$

MR. RHEE'S BRILLIANT MATH SERIES   AB & BC   AP CAL LESSON 7

For questions 7-8, find the equation of the tangent to the curve at the given point.

7. $f(x) = 2e^x + 1$ at $(1, 2)$

8. $f(x) = x^3 + 2x^2 - 2$ at $(-1, 3)$

9. For what values of x does the graph of $f(x) = \frac{1}{3}x^3 - x^2 - 3x + 1$ have a horizontal tangent?

10. Find the points on the graph of $f(x) = \sqrt[3]{x}$ where the tangent is vertical.

MR. RHEE'S BRILLIANT MATH SERIES   AB & BC   AP CAL LESSON 7

Answers

1. $f'(x) = 4x - 3$
2. $f'(x) = 2x - 3$
3. $f'(x) = -\dfrac{3}{2x\sqrt{x}}$
4. $f'(x) = -\dfrac{1}{x^2} + \dfrac{4}{x^3} - \dfrac{9}{x^4}$
5. $f'(x) = \dfrac{1}{x} + e$
6. $f'(x) = \dfrac{1}{2\sqrt{x}} + \dfrac{2}{3\sqrt[3]{x}} + \dfrac{3}{4\sqrt[4]{x}}$
7. $y - 2 = 2e(x - 1)$
8. $y - 3 = -(x + 1)$
9. $x = -1$ or $x = 3$
10. $x = 0$

MR. RHEE'S BRILLIANT MATH SERIES   AB & BC   AP CAL LESSON 8

LESSON 8

Differentiation Rules

Product Rule and Quotient Rule

Product rule and quotient rule are the differentiation rules that enable us to find the derivative of product function and quotient function.

- The Product Rule: If f and g are both differentiable, then
$$(f \cdot g)' = f' \cdot g + f \cdot g'$$

- The Quotient Rule: If If f and g are both differentiable, then
$$\left(\frac{f}{g}\right)' = \frac{f' \cdot g - f \cdot g'}{g^2}$$

[Tip]

1. Note that the derivative of a product of two functions is **NOT** the product of derivatives as shown below.
$$(f \cdot g)' \neq f' \cdot g'$$

2. The derivative of a product of three functions is as follows:
$$(f \cdot g \cdot h)' = f' \cdot g \cdot h + f \cdot g' \cdot h + f \cdot g \cdot h'$$

3. Note that the derivative of a quotient of two functions is **NOT** the quotient of derivatives as shown below.
$$\left(\frac{f}{g}\right)' \neq \frac{f'}{g'}$$

Example 1 Applying the Product rule

If $f(x) = xe^x$, find $f'(x)$.

Solution

$$f'(x) = (xe^x)' = (x)' \cdot e^x + x \cdot (e^x)'$$
$$= 1 \cdot e^x + xe^x = e^x + xe^x$$

Derivatives of Six Trigonometric Functions

$\dfrac{d}{dx}(\sin x) = \cos x$ $\qquad$ $\dfrac{d}{dx}(\cos x) = -\sin x$

$\dfrac{d}{dx}(\tan x) = \sec^2 x$ $\qquad$ $\dfrac{d}{dx}(\cot x) = -\csc^2 x$

$\dfrac{d}{dx}(\sec x) = \sec x \tan x$ $\qquad$ $\dfrac{d}{dx}(\csc x) = -\csc x \cot x$

Tip Note that all the derivatives of the trigonometric functions starting with the letter C have negative signs.

Example 2 Applying the Quotient rule

Differentiate $y = \tan x$ using the Quotient rule.

Solution Since $\tan x = \frac{\sin x}{\cos x}$,

$$\begin{aligned} y' = (\tan x)' &= \left(\frac{\sin x}{\cos x}\right)' \qquad &&\text{Apply the quotient rule} \\ &= \frac{(\sin x)' \cdot \cos x - \sin x \cdot (\cos x)'}{\cos^2 x} \\ &= \frac{\cos x \cos x - \sin x \cdot (-\sin x)}{\cos^2 x} \\ &= \frac{\cos^2 x + \sin^2 x}{\cos^2 x} \qquad &&\text{Use } \cos^2 x + \sin^2 x = 1 \\ &= \frac{1}{\cos^2 x} \\ &= \sec^2 x \end{aligned}$$

Example 3   Applying the Product rule

Find the derivative function of $y = e^x \sin x$.

Solution   Since $e^x \sin x$ is a product of two functions, apply the Product rule to the function.

$$\begin{aligned} y' &= (e^x \sin x)' \\ &= (e^x)' \cdot \sin x + e^x \cdot (\sin x)' \qquad &&\text{Since } (e^x)' = e^x \\ &= e^x \sin x + e^x \cos x \end{aligned}$$

MR. RHEE'S BRILLIANT MATH SERIES   AB & BC   AP CAL LESSON 8

EXERCISES

For questions 1-8, differentiate the following functions.

1. $y = x^2 e^x$

2. $y = \dfrac{4}{\cos x}$

3. $y = \dfrac{\sin 2x}{\cos x}$

4. $y = \sin x \cos x$

5. $y = \dfrac{\cos x + 1}{\sin x}$

6. $y = \dfrac{e^x + x}{\cos x}$

7. $y = \dfrac{\sin x + \cos x}{x}$

8. $y = xe^x \sin x$

9. Find the equation of the tangent line to the curve $y = 4\cos x$ at $\left(\dfrac{\pi}{2}, 0\right)$.

10. Find the equation of the tangent line to the curve $y = \sin x + \cos x$ at $\left(\dfrac{\pi}{4}, \sqrt{2}\right)$.

MR. RHEE'S BRILLIANT MATH SERIES AB & BC AP CAL LESSON 8

Answers

1. $y' = 2xe^x + x^2 e^x$
2. $y' = 4\sec x \tan x$
3. $y' = 2\cos x$
4. $y' = \cos^2 x - \sin^2 x$
5. $y' = -\left(\dfrac{1 + \cos x}{\sin^2 x}\right)$
6. $y' = \dfrac{(e^x + 1)\cos x + (e^x + x)\sin x}{\cos^2 x}$
7. $y' = \dfrac{(\cos x - \sin x)x - (\sin x + \cos x)}{x^2}$
8. $y' = e^x \sin x + xe^x \sin x + xe^x \cos x$
9. $y = -4\left(x - \dfrac{\pi}{2}\right)$
10. $y = \sqrt{2}$

MR. RHEE'S BRILLIANT MATH SERIES AB & BC AP CAL LESSON 9

LESSON 9

The Chain Rule

The Chain Rule

If you want to differentiate the function $f(x) = (x+1)^3$ using the differentiation rules that you have learned so far, you need to expand $(x+1)^3$ and apply the sum rule and power rule to each term of the function. Since $(x+1)^3 = x^3 + 3x^2 + 3x + 1$, the derivative function is

$$\frac{d}{dx}(x+1)^3 = (x^3 + 3x^2 + 3x + 1)' = 3x^2 + 6x + 3$$

However, a serious problem arises when you want to differentiate the function $(x+1)^{100}$. According to the Binomial theorem, $(x+1)^{100}$ can be written as

$$(x+1)^{100} = \sum_{k=0}^{100} \binom{100}{k} x^{100-k} 1^k$$

which indicates that the function $(x+1)^{100}$ has 101 terms. So, if you want to differentiate the function $f(x) = (x+1)^{100}$, you need to expand the function and differentiate 101 terms of the function. In this lesson, you will learn about the new differentiation rule called the **Chain Rule** which will help you differentiate a composition function like $f(x) = (x+1)^{100}$ with ease.

- The Chain Rule

  If f and g are both differentiable and F is the composition function defined by $F(x) = f(g(x))$, then F' is given by the product

  $$F'(x) = f'(g(x))g'(x)$$

  In Leibniz notion, if $y = f(u)$ and $y = g(x)$ are both differentiable function, then

  $$\frac{dy}{dx} = \frac{dy}{du} \cdot \frac{du}{dx}$$

Tip The Chain Rule is the one of the most important differentiation rules that you are going to use a lot throughout the AP Calculus AB and AP Calculus BC courses. Many students tend to make mistake by forgetting the Chain rule when they differentiate a composition function.

MR. RHEE'S BRILLIANT MATH SERIES AB & BC AP CAL LESSON 9

Example 1 Applying the Chain rule

Differentiate $y = (x+1)^{100}$ using the Chain rule.

Solution Let $y = u^{100}$ and $u = x+1$. If you substitute u for $x+1$, then $y = (x+1)^{100}$. Since $(x+1)^{100}$ is a composition function, apply the Chain rule to the function.

$$y = u^{100}$$
$$\frac{dy}{du} = 100u^{99}$$
$$= 100(x+1)^{99}$$

$$u = x+1$$
$$\frac{du}{dx} = 1$$

Thus,

$$\frac{dy}{dx} = \frac{dy}{du} \cdot \frac{du}{dx}$$
$$= 100(x+1)^{99} \cdot 1$$
$$= 100(x+1)^{99}$$

Therefore, the derivative function of $y = (x+1)^{100}$ is $100(x+1)^{99}$.

Example 2 Applying the Chain rule

Differentiate $y = \sqrt{1-x^2}$.

Solution Let $y = \sqrt{u}$ and $u = 1 - x^2$. If you substitute u for $1 - x^2$, then $y = \sqrt{1-x^2}$. Since $y = \sqrt{1-x^2}$ is a composition function, apply the Chain rule to the function.

$$y = \sqrt{u}$$
$$\frac{dy}{du} = \frac{1}{2\sqrt{u}}$$
$$= \frac{1}{2\sqrt{1-x^2}}$$

$$u = 1 - x^2$$
$$\frac{du}{dx} = -2x$$

Thus,

$$\frac{dy}{dx} = \frac{dy}{du} \cdot \frac{du}{dx}$$
$$= \frac{1}{2\sqrt{1-x^2}} \cdot -2x$$
$$= \frac{-x}{\sqrt{1-x^2}}$$

Therefore, the derivative function of $y = \sqrt{1-x^2}$ is $\dfrac{-x}{\sqrt{1-x^2}}$.

Example 3 Applying the Chain rule

Differentiate $y = e^{\sin x}$.

Solution Since $e^{\sin x}$ is a composition function, let $y = e^u$ and $u = \sin x$. Use the Chain rule to differentiate the function.

$$y = e^u \qquad\qquad u = \sin x$$
$$\frac{dy}{du} = e^u \qquad\qquad \frac{du}{dx} = \cos x$$
$$= e^{\sin x}$$

Thus,

$$\frac{dy}{dx} = \frac{dy}{du} \cdot \frac{du}{dx}$$
$$= e^{\sin x} \cdot \cos x$$
$$= e^{\sin x} \cos x$$

Therefore, the derivative function of $y = e^{\sin x}$ is $e^{\sin x} \cos x$.

MR. RHEE'S BRILLIANT MATH SERIES   AB & BC   AP CAL LESSON 9

EXERCISES

For questions 1-9, differentiate the following functions.

1. $y = (2x^2 + 3x)^{10}$

2. $y = \left(\dfrac{x+1}{x-1}\right)^{10}$

3. $y = \ln \sqrt{x}$

4. $y = \sin^2 x$

5. $y = \cos 4\theta$

6. $y = \dfrac{4}{\sqrt{2x-3}}$

7. $y = \sqrt[3]{1+\sin\theta}$

8. $y = \tan(\cos x)$

9. $y = \sqrt{1+\sqrt{x}}$

10. Find the equation of the tangent line to the curve $y = \sec 2x$ at $\left(\dfrac{\pi}{6}, 2\right)$.

MR. RHEE'S BRILLIANT MATH SERIES   AB & BC   AP CAL LESSON 9

Answers

1. $y' = 10(2x^2 + 3x)^9(4x + 3)$
2. $y' = -20\left(\dfrac{x+1}{x-1}\right)^9 \dfrac{1}{(x-1)^2}$
3. $y' = \dfrac{1}{2x}$
4. $y' = 2\sin x \cos x$
5. $y' = -4\sin 4\theta$
6. $y' = -\dfrac{4}{(2x-3)^{\frac{3}{2}}}$
7. $y' = \dfrac{\cos\theta}{3\sqrt[3]{(1+\sin\theta)^2}}$
8. $y' = -\sec^2(\cos x)\sin x$
9. $y' = \dfrac{1}{4\sqrt{x}\sqrt{1+\sqrt{x}}}$
10. $y - 2 = 4\sqrt{3}\left(x - \dfrac{\pi}{6}\right)$

MR. RHEE'S BRILLIANT MATH SERIES AB & BC   AP CAL LESSON 10

LESSON 10
Implicit Differentiation

Implicit Differentiation

We have learned many differentiation rules so far. These rule can be applied to a function, $y = f(x)$, defined explicitly. In case you want to differentiate some functions defined implicitly as shown below,

$$x^2 + y^2 = 25 \quad \text{(Implicit function)} \quad \Longrightarrow \quad y = \pm\sqrt{25 - x^2} \quad \text{(Explicit function)}$$

you need to redefine the functions explicitly as shown above and use the differentiation rules. However, some implicit functions like one shown below

$$x^4 + 2x^2y^2 + y^5 = 7$$

are impossible to define explicitly so that you are not able to use any one of the differentiation rules to find the derivative functions. Fortunately, the **Implicit differentiation** help you find the derivative function directly from the implicit equation without solve an equation for y in terms of x.

How to do implicit differentiation

1. Consider y as a function of x.
2. Differentiate both sides of an implicit equation with respect to x.
3. Whenever differentiating y, multiply the result by $\frac{dy}{dx}$.
4. Solve the resulting equation for $\frac{dy}{dx}$.

Tip

1. Consider y as a function of x. In other words, $y = f(x)$. So, $y^2 = [f(x)]^2$. Since $y^2 = [f(x)]^2$ is a composition function, use the Chain rule to differentiate y^2 with respect to x.

$$\begin{aligned}(y^2)' &= ([f(x)]^2)' \\ &= 2f(x) \cdot f'(x) \qquad \text{Use the Chain rule} \\ &= 2y\frac{dy}{dx} \qquad \text{Since } y = f(x) \text{ and } f'(x) = \frac{dy}{dx}\end{aligned}$$

2. In order to differentiate xy, use the Chain rule and the Product rule.

$$\begin{aligned}(xy)' &= (x)' \cdot y + x(y)' \\ &= y + x\frac{dy}{dx}\end{aligned}$$

Example 1   Applying Implicit differentiation

If $x^2 + y^2 = 25$, find $\dfrac{dy}{dx}$.

Solution   Differentiate both sides of an implicit equation with respect to x.

$$x^2 + y^2 = 25$$
$$2x + 2y\frac{dy}{dx} = 0$$
$$2y\frac{dy}{dx} = -2x$$
$$\frac{dy}{dx} = -\frac{x}{y}$$

Therefore, $\dfrac{dy}{dx} = -\dfrac{x}{y}$.

Example 2   Applying Implicit differentiation

If $x^3 + y^3 = 6xy$, find $\dfrac{dy}{dx}$.

Solution   Differentiate both sides of an implicit equation with respect to x.

$$x^3 + y^3 = 6xy \qquad \text{Use the Product rule to differentiate } xy$$
$$3x^2 + 3y^2\frac{dy}{dx} = 6\left(y + x\frac{dy}{dx}\right) \qquad \text{Since } (xy)' = y + x\frac{dy}{dx}$$
$$3x^2 + 3y^2\frac{dy}{dx} = 6y + 6x\frac{dy}{dx}$$
$$(3y^2 - 6x)\frac{dy}{dx} = 6y - 3x^2 \qquad \text{Divide both sides by } (3y^2 - 6x)$$
$$\frac{dy}{dx} = \frac{6y - 3x^2}{3y^2 - 6x}$$
$$\frac{dy}{dx} = \frac{2y - x^2}{y^2 - 2x}$$

Therefore, $\dfrac{dy}{dx} = \dfrac{2y - x^2}{y^2 - 2x}$.

MR. RHEE'S BRILLIANT MATH SERIES — AB & BC — AP CAL LESSON 10

Example 3 Applying Implicit differentiation

If $\sqrt{xy} = \sqrt{x+y}$, find $\dfrac{dy}{dx}$.

Solution Differentiate both sides of an implicit equation with respect to x.

$$\sqrt{xy} = \sqrt{x+y}$$

$$\frac{1}{2\sqrt{xy}}(xy)' = \frac{1}{2\sqrt{x+y}}(x+y)'$$

$$\frac{1}{2\sqrt{xy}}\left(y + x\frac{dy}{dx}\right) = \frac{1}{2\sqrt{x+y}}\left(1 + \frac{dy}{dx}\right)$$

$$\frac{y}{2\sqrt{xy}} + \frac{x}{2\sqrt{xy}}\frac{dy}{dx} = \frac{1}{2\sqrt{x+y}} + \frac{1}{2\sqrt{x+y}}\frac{dy}{dx}$$

$$\left(\frac{x}{2\sqrt{xy}} - \frac{1}{2\sqrt{x+y}}\right)\frac{dy}{dx} = \frac{1}{2\sqrt{x+y}} - \frac{y}{2\sqrt{xy}}$$

$$\frac{dy}{dx} = \frac{\frac{1}{2\sqrt{x+y}} - \frac{y}{2\sqrt{xy}}}{\frac{x}{2\sqrt{xy}} - \frac{1}{2\sqrt{x+y}}}$$

MR. RHEE'S BRILLIANT MATH SERIES — AB & BC — AP CAL LESSON 10

EXERCISES

1. Find $\dfrac{dy}{dx}$ if $\dfrac{1}{x} + \dfrac{1}{y} = 4$.

2. Find $\dfrac{dy}{dx}$ if $\sqrt{x} + \sqrt{y} = 1$.

3. Find $\dfrac{dy}{dx}$ if $x^2 y + xy^2 = 3$.

4. Find $\dfrac{dy}{dx}$ if $\dfrac{1}{y} + xy = 5$.

5. Find $\dfrac{dy}{dx}$ if $\sin x \cos y = 2$.

MR. RHEE'S BRILLIANT MATH SERIES	AB & BC	AP CAL LESSON 10

6. Find $\dfrac{dy}{dx}$ if $\cos(x+y) = xe^x$.

7. Find the equation of the tangent line to the curve $x^2 + y^2 = 25$ at $(-3, 4)$.

8. Find the equation of the tangent line to the curve $x^3 + y^3 = 6xy$ at $(3, 3)$.

Answers

1. $\dfrac{dy}{dx} = -\dfrac{y^2}{x^2}$

2. $\dfrac{dy}{dx} = -\dfrac{\sqrt{y}}{\sqrt{x}}$

3. $\dfrac{dy}{dx} = -\left(\dfrac{2xy + y^2}{x^2 + 2xy}\right)$

4. $\dfrac{dy}{dx} = -\left(\dfrac{y}{x - \frac{1}{y^2}}\right)$

5. $\dfrac{dy}{dx} = \dfrac{\cos x \cos y}{\sin x \sin y}$

6. $\dfrac{dy}{dx} = -\left(\dfrac{e^x + xe^x + \sin(x+y)}{\sin(x+y)}\right)$

7. $y - 4 = \dfrac{3}{4}(x + 3)$

8. $y - 3 = -(x - 3)$

MR. RHEE'S BRILLIANT MATH SERIES   AB & BC   AP CAL LESSON 11

LESSON 11

Derivatives of Inverse Trig Functions and Higher Derivatives

Derivatives of Inverse Trigonometric Functions

$$\frac{d}{dx}(\sin^{-1} x) = \frac{1}{\sqrt{1-x^2}} \qquad \frac{d}{dx}(\csc^{-1} x) = -\frac{1}{x\sqrt{x^2-1}}$$

$$\frac{d}{dx}(\cos^{-1} x) = -\frac{1}{\sqrt{1-x^2}} \qquad \frac{d}{dx}(\sec^{-1} x) = \frac{1}{x\sqrt{x^2-1}}$$

$$\frac{d}{dx}(\tan^{-1} x) = \frac{1}{1+x^2} \qquad \frac{d}{dx}(\cot^{-1} x) = -\frac{1}{1+x^2}$$

Tip Most likely the derivatives of the $\sin^{-1} x$ and $\tan^{-1} x$ will be on the AP Calculus exam.

Example 1 Finding the derivative of Inverse tangent

Find $\dfrac{dy}{dx}$ if $y = \tan^{-1} 4x$.

Solution Let $y = \tan^{-1} u$ and $u = 4x$. Since $y = \tan^{-1} 4x$ is a composition function, use the Chain rule to find $\dfrac{dy}{dx}$.

$$y = \tan^{-1} u \qquad\qquad u = 4x$$
$$\frac{dy}{du} = \frac{1}{1+u^2} \qquad\qquad \frac{du}{dx} = 4$$
$$= \frac{1}{1+(4x)^2}$$

$$\frac{dy}{dx} = \frac{dy}{du} \cdot \frac{du}{dx}$$
$$= \frac{1}{1+(4x)^2} \cdot 4$$
$$= \frac{1}{1+16x^2}$$

Therefore, the derivative function of $y = \tan^{-1} 4x$ is $\dfrac{4}{1+16x^2}$.

Higher Derivatives

If f is a differentiable function, the derivative of f is called the **first derivative**. The notations of the first derivative are

$$f' = y' = \frac{dy}{dx}$$

The derivative of the first derivative is called **second derivative** which is denoted by

$$f'' = y'' = \frac{d^2y}{dx^2}$$

The **third derivative** is the derivative of second derivative and is denoted by

$$f''' = y''' = \frac{d^3y}{dx^3}$$

In general, the nth derivative of f is obtained by differentiating f n times and is denoted by

$$f^{(n)}(x) = y^{(n)} = \frac{d^n y}{dx^n}$$

MR. RHEE'S BRILLIANT MATH SERIES
AB & BC — AP CAL LESSON 11

> **Velocity and Acceleration**
>
> If $f(t)$ is the position function of a particle, the first derivative of $f(t)$ represents **velocity** and is denoted by $v(t) = f'(t)$. The second derivative of the position function is called **acceleration** and is the derivative of the velocity. The acceleration is denoted by $a(t) = v'(t) = f''(t)$.
>
> The following guidelines will help you solve a problem regarding velocity and acceleration.
>
> 1. Time at which a particle is at rest: Let v(t)=0 and solve for t.
>
> 2. Time at which a particle about to change its direction: Let v(t)=0 and solve for t.
>
> 3. When the particle speed up or slow down:
>
>   - The particle speed up: when the velocity and acceleration have the same sign.
>   - The particle slow down: when the velocity and acceleration have the opposite signs.

Example 2 Finding the velocity and acceleration

If the position function $f(t)$ is defined by $f(t) = t^3 - 3t^2 - 9t$, where t measured in seconds and f in meters.

(a) Find the velocity at $t = 3$.

(b) Find the acceleration at $t = 2$

Solution

(a) Since the velocity function is the first derivative of position function,
$$v(t) = f'(t) = 3t^2 - 6t - 9$$

Thus, the velocity at $t = 3$ is $v(3) = 3(3)^2 - 6(3) - 9 = 0$, which indicates that the particle is at rest or is about to change its direction.

(b) Since the acceleration function is the first derivative of velocity function,
$$a(t) = v'(t) = 6t - 6$$

Thus, the acceleration at $t = 2$ is $a(2) = 6\ m/s^2$.

Example 3 Finding the first and second derivatives

If $f(x) = xe^x$, find $f'(x)$ and $f''(x)$.

Solution Since xe^x is the product of two function, use the Product rule to differentiate xe^x with respect to x.

$$f'(x) = (x \cdot e^x)' = e^x + xe^x$$

Since $f''(x)$ is the derivative of $f'(x)$, differentiate $f'(x)$ with respect to x again.

$$\begin{aligned} f''(x) &= (e^x + xe^x)' \\ &= (e^x)' + (xe^x)' \\ &= e^x + e^x + xe^x \\ &= 2e^x + xe^x \end{aligned}$$

MR. RHEE'S BRILLIANT MATH SERIES — AB & BC — AP CAL LESSON 11

EXERCISES

1. If $y = \sqrt{1-x^2}\,\sin^{-1} x$, find y'.

2. If $y = \tan^{-1}\left(\frac{2}{x}\right)$, find $\frac{dy}{dx}$.

3. If $y = \sin 2x$, find y''.

4. If $f(x) = \dfrac{x}{1-x}$, find $f''(x)$.

5. If $x^2 + y^2 = 9$, find $\dfrac{dy}{dx}$ and $\dfrac{d^2y}{dx^2}$.

6. A particle moves along the x-axis. The position of particle at time t given by
$$s(t) = \frac{1}{3}t^3 - 2t^2 + 3t, \qquad 0 \leq t < 5$$
where t is measured in seconds and s in feet.

   (a) Find the acceleration at $t = 4$.

   (b) Find the time at which the particle changes its direction.

   (c) When is the particle speeding up?

MR. RHEE'S BRILLIANT MATH SERIES — AB & BC — AP CAL LESSON 11

Answers

1. $y' = -\dfrac{x\sin^{-1} x}{\sqrt{1-x^2}} + 1$

2. $\dfrac{dy}{dx} = -\dfrac{2}{x^2+4}$

3. $y'' = -4\sin 2x$

4. $f''(x) = \dfrac{2}{(1-x)^3}$

5. $\dfrac{dy}{dx} = -\dfrac{x}{y}$, $\qquad \dfrac{d^2y}{dx^2} = -\left(\dfrac{y^2+x^2}{y^3}\right)$

6 (a). $a(4) = 4$

6 (b). At $t=1$ and $t=3$

6 (c). $1 < t < 2$ and $3 < t < 5$

LESSON 12

Indeterminate Forms And L'Hospital's Rule

Indeterminate Forms

Suppose you are trying to find the limit of the following function algebraically.

$$\lim_{x \to 0} \frac{\sin x}{x}$$

The first step you should do is to substitute x for 0 in the numerator and the denominator. As you notice, $\sin x \to 0$ as $x \to 0$. So, the limit you are trying to find has a form of $\frac{0}{0}$. In general, $\frac{0}{0}$ and $\frac{\infty}{\infty}$ are called **Indeterminate forms** for which the limit may or may not exist.

Finding the limit of the indeterminate forms of $\frac{0}{0}$ and $\frac{\infty}{\infty}$ is not easy. So, mathematicians developed a method known as **L'Hospital's Rule** to find the limit of the indeterminate forms of $\frac{0}{0}$ and $\frac{\infty}{\infty}$ easily.

(Tip)

1. Indeterminate form of $\frac{0}{0}$ is common in Calculus because the derivative has an indeterminate form of $\frac{0}{0}$. From the definition of the derivative shown below, substitute h for 0 in the numerator and the denominator.

$$f'(x) = \lim_{h \to 0} \frac{f(x+h) - f(x)}{h} = \frac{f(x) - f(x)}{0} = \frac{0}{0}$$

You will get a $\frac{0}{0}$.

2. You expect to solve a lot of problems regarding the limit of an indeterminate form of $\frac{0}{0}$ due to the fact that the importance of the indeterminate form of $\frac{0}{0}$ is pretty high.

MR. RHEE'S BRILLIANT MATH SERIES — AB & BC — AP CAL LESSON 12

L'Hospital's Rule

Let f and g are differentiable and $g'(x) \neq 0$ near a. Suppose you have one of the following cases

$$\lim_{x \to a} \frac{f(x)}{g(x)} = \frac{0}{0} \qquad \text{or} \qquad \lim_{x \to a} \frac{f(x)}{g(x)} = \frac{\pm\infty}{\pm\infty}$$

where a can be any real number, infinity, or negative infinity. Then,

$$\lim_{x \to a} \frac{f(x)}{g(x)} = \lim_{x \to a} \frac{f'(x)}{g'(x)}$$

1. L'Hospital's Rule says that the limit of a quotient of functions is equal to the limit of the quotient of their derivatives.

2. L'Hospital's Rule is also valid for one-sided limits: that is, $x \to a$ can be replaced by any of the followings: $x \to a^+$, $x \to a^-$.

Example 1 Finding the limit

Find $\lim\limits_{x \to -2} \dfrac{x^3 + 8}{x + 2}$.

Solution Substitute x for -2 in the numerator and the denominator. Then, you will get an indeterminate form of $\dfrac{0}{0}$. Use the L'Hospital's rule to find the limit.

$$\begin{aligned}
\lim_{x \to -2} \frac{x^3 + 8}{x + 2} &= \lim_{x \to -2} \frac{(x^3 + 8)'}{(x + 2)'} \\
&= \lim_{x \to -2} \frac{3x^2}{1} \qquad \text{Substitute } x \text{ for } -2 \\
&= 12
\end{aligned}$$

Therefore, $\lim\limits_{x \to -2} \dfrac{x^3 + 8}{x + 2} = 12$.

MR. RHEE'S BRILLIANT MATH SERIES — AB & BC — AP CAL LESSON 12

Example 2 Finding the limit

Find $\lim\limits_{x \to 0} \dfrac{\sin x}{x}$.

Solution Substitute x for 0 in the numerator and the denominator. Then, you will get an indeterminate form of $\dfrac{0}{0}$. Use the L'Hospital's rule to find the limit.

$$\begin{aligned}
\lim_{x \to 0} \frac{\sin x}{x} &= \lim_{x \to 0} \frac{(\sin x)'}{(x)'} \\
&= \lim_{x \to 0} \frac{\cos x}{1} \qquad \text{Substitute } x \text{ for } 0 \\
&= 1
\end{aligned}$$

Therefore, $\lim\limits_{x \to 0} \dfrac{\sin x}{x} = 1$.

Example 3 Finding the limit

Find $\lim\limits_{x \to 2} \dfrac{x-2}{\ln(x-1)}$.

Solution Substitute x for 2 in the numerator and the denominator. Then, you will get an indeterminate form of $\dfrac{0}{0}$. Use the L'Hospital's rule to find the limit.

$$\begin{aligned}
\lim_{x \to 2} \frac{x-2}{\ln(x-1)} &= \lim_{x \to 2} \frac{(x-2)'}{(\ln(x-1))'} \\
&= \lim_{x \to 2} \frac{1}{\frac{1}{x-1}} \qquad \text{Use the Chain rule to differentiate } \ln(x-1) \\
&= \lim_{x \to 2} (x-1) \\
&= 1
\end{aligned}$$

Therefore, $\lim\limits_{x \to 2} \dfrac{x-2}{\ln(x-1)} = 1$.

MR. RHEE'S BRILLIANT MATH SERIES — AB & BC — AP CAL LESSON 12

Example 4 Finding the limit

Find $\lim\limits_{x \to \infty} \dfrac{e^x}{x^2}$.

Solution Substitute x for ∞ in the numerator and the denominator. Then, you will get an indeterminate form of $\dfrac{\infty}{\infty}$. Use the L'Hospital's rule to find the limit.

$$\begin{aligned}
\lim_{x \to \infty} \frac{e^x}{x^2} &= \lim_{x \to \infty} \frac{(e^x)'}{(x^2)'} \\
&= \lim_{x \to \infty} \frac{e^x}{2x} &&\text{Substitute } x \text{ for } \infty, \text{ you get a } \frac{\infty}{\infty} \\
&= \lim_{x \to \infty} \frac{(e^x)'}{(2x)'} &&\text{Use the L'Hospital rule again} \\
&= \lim_{x \to \infty} \frac{e^x}{2} &&\text{Substitute } x \text{ for } \infty \text{ again} \\
&= \frac{\infty}{2} \\
&= \infty
\end{aligned}$$

Therefore, $\lim\limits_{x \to \infty} \dfrac{e^x}{x^2} = \infty$.

EXERCISES

1. Find $\lim\limits_{x \to 0} \dfrac{\sin 2x}{3x}$.

2. Find $\lim\limits_{x \to \infty} \dfrac{\ln 2x}{x}$.

3. Find $\lim\limits_{x \to 1} \dfrac{\sqrt[3]{x} - 1}{x - 1}$.

4. Find $\lim\limits_{x \to 0} \dfrac{\tan 3x}{5x}$.

5. Find $\lim\limits_{x \to \infty} \dfrac{e^x}{x^3}$.

MR. RHEE'S BRILLIANT MATH SERIES   AB & BC   AP CAL LESSON 12

6. Find $\lim\limits_{x \to 0} \dfrac{\sin x}{e^x - 1}$.

7. Find $\lim\limits_{x \to 0} \dfrac{\sin^{-1} x}{x}$.

8. Find $\lim\limits_{x \to 0} \dfrac{1 - \cos 4x}{3x}$.

MR. RHEE'S BRILLIANT MATH SERIES AB & BC AP CAL LESSON 12

Answers

1. $\dfrac{2}{3}$ 2. 0
3. $\dfrac{1}{3}$ 4. $\dfrac{3}{5}$
5. ∞ 6. 1
7. 1 8. 0

MR. RHEE'S BRILLIANT MATH SERIES AB & BC AP CAL LESSON 13

LESSON 13

Related Rates

Related Rates

In a related rates problem, we are trying to find the rate of change of one quantity in terms of the rate of change of another quantity that is already measured. Related rates is an application of an implicit differentiation. The quantities that you are about to see in a related rate problem are functions of time t.

How to do related rates problems

1. Read a problem and draw a diagram.
2. Identify all quantities that are function of time t.
3. Assign variables to all quantities that are function of time t.
4. Write down all information and given rates in terms of derivatives.
5. Set up an equation that relates the quantities.
6. Use the Chain rule to differentiate both sides of the equation with respect to t.
7. Solve for the unknown rate.

Tip — Suppose x is a quantity that is a function of time t in a related rates problem: $x = f(t)$. Below shows you how to differentiate x^2 with respect to t using the Chain rule.

$$(x^2)' = \frac{d}{dt}(f(t))^2 \qquad \text{Since } x = f(t), \text{ Use the Chain rule}$$

$$= 2f(t) \cdot f'(t) \qquad \text{Since } f(t) = x \text{ and } f'(t) = \frac{dx}{dt}$$

$$= 2x\frac{dx}{dt}$$

Example 1 Solving a related rates problem

A ladder 10 ft long leans against a building. If the bottom of the ladder slides away from the wall at a rate of 2 ft/sec, how fast is the top of the ladder sliding down the wall when the bottom of the ladder is 6 ft from the wall?

Solution Let x be the distance from the bottom of the ladder to the wall and y be the distance from the top of the ladder to the ground as shown in figures below. Since the top of the ladder slides down, y decreases as time increases. However, the bottom of the lader slides away. So, x increases as time increases. Thus, x and y are functions of time t.

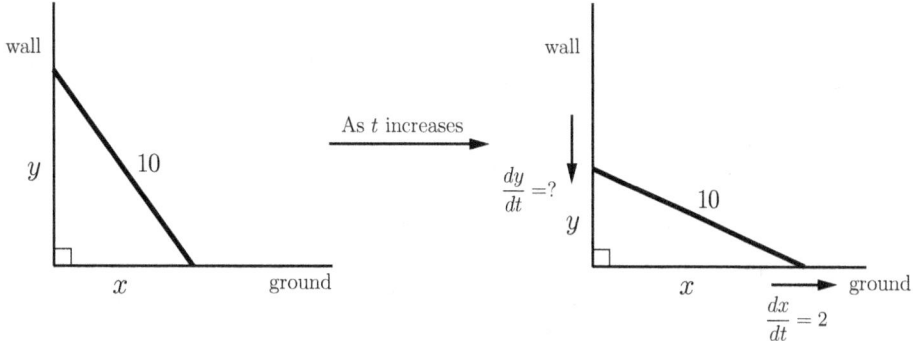

We are given that $\dfrac{dx}{dt} = 2$ ft/sec and try to find $\dfrac{dy}{dt}$ when $x = 6$ ft. Let's set up an equation that relates x and y using the Pythagorean theorem.

$$x^2 + y^2 = 10^2$$

Differentiate both sides with respect to t using the Chain rule and solve for $\dfrac{dy}{dt}$.

$$2x\frac{dx}{dt} + 2y\frac{dy}{dt} = 0$$
$$2y\frac{dy}{dt} = -2x\frac{dx}{dt}$$
$$\frac{dy}{dt} = -\frac{x}{y}\frac{dx}{dt}$$

From the equation $x^2 + y^2 = 10^2$, $y = 8$ when $x = 6$. Substituting $x = 6$, $y = 8$ and $\dfrac{dx}{dt} = 2$,

$$\frac{dy}{dt} = -\frac{x}{y}\frac{dx}{dt} = -\frac{6}{8}(2) = -\frac{3}{2}$$

Therefore, the top of the ladder is sliding down the wall at a rate of $-\frac{3}{2}$ ft/sec.

MR. RHEE'S BRILLIANT MATH SERIES — AB & BC — AP CAL LESSON 13

Example 2 Solving a related rates problem

If a spherical snowball melts so that its volume decreases at a rate of 1 cm³/min, find the rate at which the radius decreases when the radius is 10 cm.

Solution Let V be the volume of the sphere and r be the radius of the sphere. As shown in the figures below, both the volume and radius decrease as time t increases. So, the volume and the radius are the functions of t.

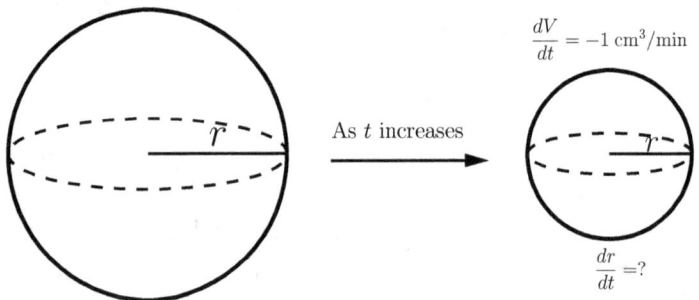

We are given that $\dfrac{dV}{dt} = -1$ cm³/min and try to find $\dfrac{dr}{dt}$ when $r = 10$ cm. Let's set up an equation that relates V and r.

$$V = \frac{4}{3}\pi r^3$$

Differentiate both sides with respect to t using the Chain rule and solve for $\dfrac{dr}{dt}$.

$$\frac{dV}{dt} = 4\pi r^2 \frac{dr}{dt}$$

$$\frac{dr}{dt} = \frac{1}{4\pi r^2} \frac{dV}{dt}$$

Substituting $r = 10$, $\dfrac{dV}{dt} = -1$

$$\begin{aligned}\frac{dr}{dt} &= \frac{1}{4\pi r^2} \frac{dV}{dt} \\ &= \frac{1}{4\pi(10)^2}(-1) \\ &\approx -0.000796\end{aligned}$$

Therefore, the rate at which the radius decreases when the radius is 10 cm is 0.000796 cm/min

MR. RHEE'S BRILLIANT MATH SERIES AB & BC AP CAL LESSON 13

EXERCISES

1. An airplane flying horizontally at 600 mph at an altitude of 2 miles passes directly over a radar station. Find the rate at which the distance from the airplane and the station is increasing when it is 4 miles away from the station.

2. Two trains start moving from the same station at noon. One travels north at 50 mph and the other travels west at 60 mph. At what rate is the distance between the trains increasing at 3 P.M. ?

3. A man walks along a straight path at a speed of 6 ft/s. A searchlight is located on the ground 30 ft from the path and is kept focused on the man. At what rate is the searchlight rotating when the man is 40 ft from the point on the path closest to the searchlight?

MR. RHEE'S BRILLIANT MATH SERIES — AB & BC — AP CAL LESSON 13

4. A water tank has the shape of an inverted circular cone with base radius 4 ft and height 10 ft. Water is being pumped into the tank at a rate of 6 ft^3/ min.

   (a) Find the rate at which the radius of the water surface is increasing when the water is 4 ft deep.

   (b) Find the rate at which the water level is rising when the water is 4 ft deep.

Answers

1. $300\sqrt{3}$ mph
2. 78.1 mph
3. 0.072 rad/sec
4(a). 0.298 ft/min
4(b). 0.745 ft/min

MR. RHEE'S BRILLIANT MATH SERIES   AB & BC   AP CAL LESSON 14

LESSON 14

Linear Approximations And Differentials

Linear Approximations

A curve f lies very close to its tangent line near $x = a$ as shown in Figure 1. So, we can approximate a value of f using the tangent line near $x = a$. An equation of the tangent line is

$$L(x) = f(a) + f'(a)(x - a)$$

and is called **Linear Approximation** of f at a.

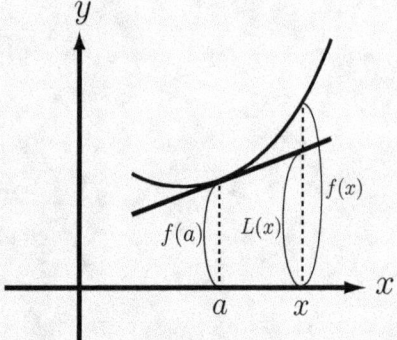
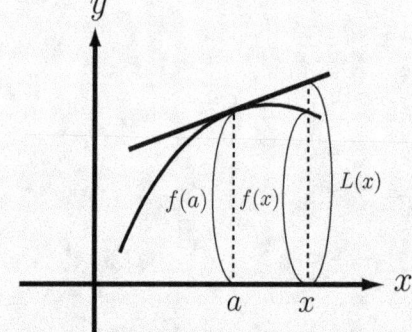

Figure 1   Figure 2

Depending on how the curve f lies either above or below the tangent line, linear approximation will be either an underestimate or an overestimate.

1. In Figure 1, $f(x)$ lies above the tangent line. In other words, $f(x) > L(x)$. Thus, $L(x)$ is an underestimate.

2. In Figure 2, $f(x)$ lies below the tangent line. In other words, $f(x) < L(x)$. Thus, $L(x)$ is an overestimate.

 Tip   When you do a linear approximation, choose the value of a so that $f(a)$ is easy to calculate by hand. In general, $f(a)$ is easy to calculate but, nearby values of f, $f(x)$ in Figure 1, is difficult to calculate by hand.

Example 1 Finding the linear approximation of the function

Find the linear approximation of the function $f(x) = \sqrt{x-1}$ at $a = 5$ and use it to approximate the value $\sqrt{5}$. Determine whether the approximation is an overestimate or an underestimate.

Solution Since $f(x) = \sqrt{x-1}$ is a composition function, use the Chain rule to find the derivative of $f(x)$.

$$f'(x) = \frac{1}{2}(x-1)^{-\frac{1}{2}}(x-1)' = \frac{1}{2\sqrt{x-1}}$$

So, $f'(5) = \frac{1}{4}$ and $f(5) = 2$. Substituting these value into $L(x) = f(a) + f'(a)(x-a)$ where $a = 5$,

$$L(x) = f(5) + f'(5)(x-5)$$
$$= 2 + \frac{1}{4}(x-5)$$
$$= \frac{1}{4}x + \frac{3}{4}$$

The linear approximation of $f(x)$ is $L(x) = \frac{1}{4}x + \frac{3}{4}$. Since $f(6) = \sqrt{5}$, you need to substitute x for 6 in $L(x)$. to approximate the value $\sqrt{5}$.

$$f(x) \approx L(x) = \frac{1}{4}x + \frac{3}{4}$$
$$\sqrt{5} = f(6) \approx L(6) = \frac{1}{4}(6) + \frac{3}{4} = \frac{9}{4} = 2.25$$

Thus, the $\sqrt{5} = 2.25$ using the linear approximation of $f(x) = \sqrt{x-1}$ at $a = 5$.

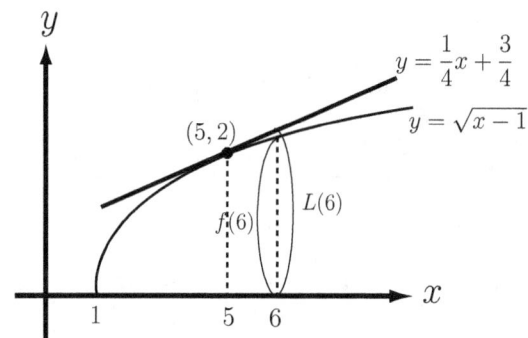

The approximation is an overestimate because the tangent line $L(x) = \frac{1}{4}x + \frac{3}{4}$ lies above the curve $f(x) = \sqrt{x-1}$ as shown in the figure above.

MR. RHEE'S BRILLIANT MATH SERIES — AB & BC — AP CAL LESSON 14

Differentials

In Figure 3 and 4, dx is called the **differential** dx and it represents very small amount in x. Similarly, dy is called the **differential** dy and it represents very small amount in y.

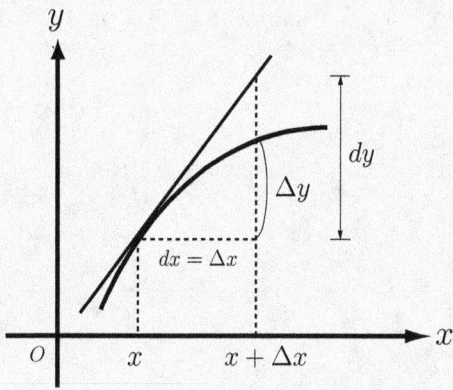

Figure 3

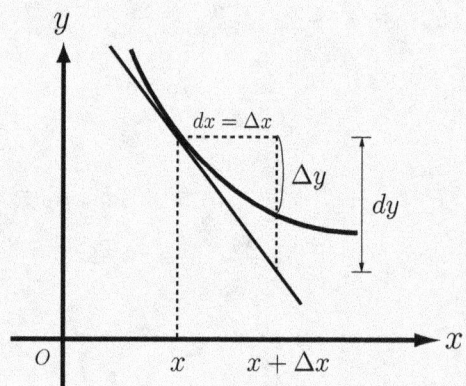

Figure 4

The slope of the tangent line at x in terms of dx and dy is

$$f'(x) = \frac{dy}{dx}$$

and multiply both sides by dx, dy can be redefined as

$$dy = f'(x)dx$$

The geometric meanings of dy and Δy are as follows:
dy represents the amount that the tangent line rises or falls when x changes by the amount dx. Whereas, Δy represents the amount that the curve $y = f(x)$ rises or fall when x changes by the amount dx as shown in Figure 3 and Figure 4.

1. Consider dx and Δx are the same in Figure 3 and Figure 4.

2. The differentials $dy = f'(x)dx$ will play an important role when you learn about the integration by the Substitution rule.

Example 2 Evaluating dy and Δy

Let $f(x) = x^2 + 1$.

(a) Find the differential dy.

(b) Evaluate dy and Δy as x changes from 1 to 1.2.

Solution

(a) Let's find the derivative of f first.
$$f'(x) = (x^2 + 1)' = 2x$$

Thus, $dy = f'(x)dx = 2xdx$.

(b) Since x changes from 1 to 1.2, $dx = \Delta x = 1.2 - 1 = 0.2$. Substituting $dx = 0.2$ and $x = 1$ in dy,
$$dy = f'(x)dx = f'(1)(0.2) = 2(1)(0.2) = 0.4$$

Whereas,
$$\Delta y = f(1.2) - f(1) = (1.2^2 + 1) - (1^2 + 1) = 0.44$$

Thus, $dy = 0.4$ and $\Delta y = 0.44$.

EXERCISES

For questions 1-4, find the linear approximation of f at given a.

1. $f(x) = e^{-x}$ at $a = 0$.

2. $f(x) = \ln x$ at $a = 1$.

3. $f(x) = \dfrac{1}{x-2}$ at $a = 3$.

4. $f(x) = \tan x$ at $a = \dfrac{\pi}{3}$.

MR. RHEE'S BRILLIANT MATH SERIES — AB & BC — AP CAL LESSON 14

For questions 5-8, find the differential of the following functions.

5. $y = x^3 + 3x + 1$

6. $y = xe^x$

7. $y = \sqrt{1 + x^2}$

8. $y = \cos 2x$

9. If $f(x) = \sqrt{x + 4}$, evaluate dy and Δy as x changes from 5 to 5.3.

10. The length of a cube is measured as 10 cm with a maximum error in measurement of 0.2 cm. Find the maximum error in the volume of the cube.

MR. RHEE'S BRILLIANT MATH SERIES AB & BC   AP CAL LESSON 14

Answers

1. $L(x) = 1 - x$
2. $L(x) = x - 1$
3. $L(x) = -x + 4$
4. $L(x) = \sqrt{3} + 4\left(x - \dfrac{\pi}{3}\right)$
5. $dy = (3x^2 + 3)\,dx$
6. $dy = (e^x + xe^x)\,dx$
7. $dy = \dfrac{x}{\sqrt{1+x^2}}\,dx$
8. $dy = -2\sin 2x\,dx$
9. $dy = 0.05, \quad \Delta y = 0.0496$
10. $dV = 60\,\text{cm}^3$

LESSON 15

Maximum And Minimum Values

Absolute Extrema and Local Extrema

A function f has an **absolute maximum** (or global maximum) at c if $f(c) \geq f(x)$ for all x in the entire domain. Whereas, f has an **absolute minimum** (or global minimum) at c if $f(c) \leq f(x)$ for all x in the entire domain. The absolute maximum and absolute minimum are called the **absolute exrema**. Suppose f is continuous on a closed interval $[a, b]$ as shown in Figure 1. f has an absolute maximum at $x = r$ and an absolute minimum at $x = s$ in the entire domain $[a, b]$.

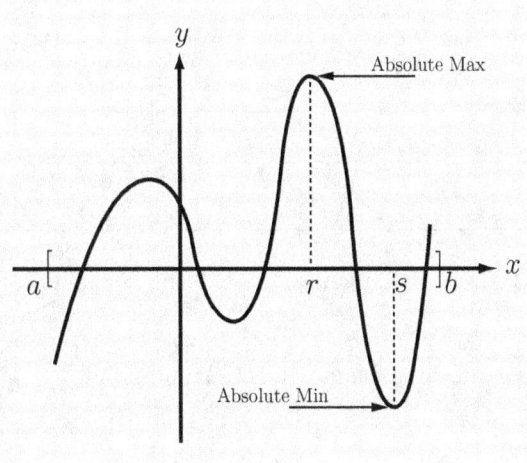

Figure 1

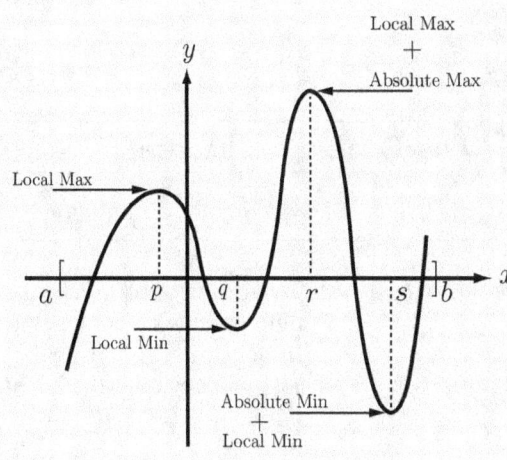

Figure 2

Similarly, f has a **local maximum** (or relative maximum) at c if $f(c) \geq f(x)$ for all x in the part of the domain, not in the entire domain. Whereas, f has a **local minimum** (or relative minimum) at c if $f(c) \leq f(x)$ for all x in the part of the domain. The local maximum and local minimum are called the **local exrema**. In Figure 2, f is continuous on a closed interval $[a, b]$. f has a local maximum at $x = p$, a local minimum at $x = q$, a local maximum at $x = r$, and a local minimum at $x = s$ in the part of the domain.

 1. Note that a local maximum can be an absolute maximum. Similarly, a local minimum can be an absolute minimum as shown in Figure 2. It depends on the graph of f.

2. A closed interval $[a, b]$ means $a \leq x \leq b$. Whereas, an open interval (a, b) means $a < x < b$.

MR. RHEE'S BRILLIANT MATH SERIES — AB & BC — AP CAL LESSON 15

The Extreme Value Theorem

If f is continuous on a closed interval $[a,b]$, then f has an absolute maximum at c and an absolute minimum at d where c and d in $[a,b]$.

 1. Some functions have extrema, whereas others do not. The Extreme value theorem gives conditions under which a function is guaranteed to have absolute extrema.

2. If f is not continuous on a closed interval $[a,b]$, the Extreme value theorem does not guarantee that f have absolute extrema.

The Fermat's Theorem

If f has a local maximum or minimum at c, and if $f'(c)$ exists, then $f'(c) = 0$.

 The converse of Fermat's theorem: If $f'(c) = 0$, then f has a local maximum or minimum at c.
In general, the converse of Fermat's theorem is false. For instance, if $f(x) = x^3$, $f'(x) = 3x^2$. So, $f'(0) = 0$. However, f does not have a local maximum or a local minimum at $x = 0$.

Definition of a Critical Number

A critical number of a function f is a number c in the domain of f such that $f'(c) = 0$ or $f'(c)$ does not exist.

 1. The Fermat's theorem can be rephrased in terms of a critical number:
If f has a local maximum or minimum at c, then c is a critical number of f.

2. The critical number is important when you find a local maximum or a local minimum because f might have a local maximum or a local minimum at the critical number.

MR. RHEE'S BRILLIANT MATH SERIES — AB & BC — AP CAL LESSON 15

Example 1 Finding a critical number

If $f(x) = x^3 - 3x^2 - 9x$, find a critical number of f.

Solution Find the derivative of f. So, $f'(x) = 3x^2 - 6x - 9$. Set the derivative of f equal to 0 and solve for x.

$$3x^2 - 6x - 9 = 0$$
$$3(x^2 - 2x - 3) = 0$$
$$3(x+1)(x-3) = 0$$
$$x = -1 \quad \text{or} \quad x = 3$$

Therefore, the critical number of f are -1 and 3.

The Closed Interval Method

In order to find the absolute maximum and minimum values of a continuous function f on a closed interval $[a, b]$, use the Closed Interval method shown below:

1. Find the critical numbers of f at c, where c in (a, b) and evaluate $f(c)$.

2. Find the function's values at the endpoints of the interval; that is, $f(a)$ and $f(b)$.

3. Comparing $f(c)$, $f(a)$, and $f(b)$. The largest value is the absolute maximum and the smallest value is the absolute minimum.

Tip
1. In step 1 in the Closed Interval method, $f(c)$ might be local extrema (local maximum or minimum) or a absolute extrema (absolute maximum or minimum).

2. In step 2 in the Closed Interval method, $f(a)$ and $f(b)$ can be only the absolute maximum or absolute minimum according to the Extreme Value theorem. In other words, $f(a)$ and $f(b)$ **CANNOT** be a local maximum or a local minimum.

MR. RHEE'S BRILLIANT MATH SERIES — AB & BC — AP CAL LESSON 15

Example 2 Finding the absolute maximum and absolute minimum

If $f(x) = 2x^3 + 3x^2 - 12x$, find the absolute maximum and absolute minimum on the closed interval $[0, 3]$.

Solution Since $f'(x) = 6x^2 + 6x - 12 = 6(x-1)(x+2)$, $f'(x) = 0$ when $x = 1$ or $x = -2$. So, the critical numbers are 1 and -2. Evaluate $f(1)$ and $f(-2)$.

$$f(1) = 2(1)^3 + 3(1)^2 - 12(1) = -7$$
$$f(-2) = 2(-2)^3 + 3(-2)^2 - 12(-2) = 20$$

Evaluate the function's values at the endpoints of the interval: $f(0) = 0$ and $f(3) = 45$. Comparing $f(1)$, $f(-2)$, $f(0)$, and $f(3)$, the absolute maximum is $f(3) = 45$, and the absolute minimum is $f(1) = -7$.

MR. RHEE'S BRILLIANT MATH SERIES — AB & BC — AP CAL LESSON 15

EXERCISES

For questions 1-5, find the critical numbers of the functions.

1. $f(x) = x^3 - 3x^2 + 4$

2. $f(x) = xe^x$

3. $f(x) = \dfrac{x}{x+1}$

4. $f(x) = x \ln x$

5. $f(x) = x\sqrt{x-1}$

For questions 6-8, find the absolute maximum and absolute minimum of f on the given interval.

6. $f(x) = \sin x + \cos x, \quad [0, \frac{\pi}{3}]$

7. $f(x) = \frac{2}{3}x^3 + 3x^2 - 8x, \quad [-5, 3]$

8. $f(x) = x^4 - 4x^2 + 3, \quad [-1, 4]$

MR. RHEE'S BRILLIANT MATH SERIES — AB & BC — AP CAL LESSON 15

Answers

1. 0, 2
2. -1
3. -1
4. $\dfrac{1}{e}$
5. 1, $\dfrac{2}{3}$
6. Absolute maximum $= 1.414$, Absolute minimum $= 1$
7. Absolute maximum $= 37.33$, Absolute minimum $= -1$
8. Absolute maximum $= 195$, Absolute minimum $= -1$

LESSON 16

The Mean Value Theorem And Rolle's Theorem

The Mean Value Theorem

Let f be a function that satisfies the following hypotheses:

1. f is continuous on the closed interval $[a,b]$.

2. f is differentiable on the open interval (a,b).

Then there is a c is (a,b) such that

$$f'(c) = \frac{f(b) - f(a)}{b - a}$$

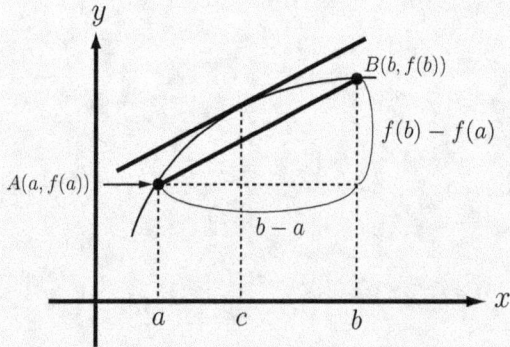

The Mean Value theorem can be stated in terms of slopes. $\dfrac{f(b) - f(a)}{b - a}$ represents the slope of the line that passes through the points A and B, and $f'(c)$ represents the slope of the tangent line at $x = c$ as shown in the figure above. Thus, the conclusion of the Mean Value theorem is that there is c in (a,b) such that the tangent line at $x = c$ is parallel to the line that passes through the points A and B.

MR. RHEE'S BRILLIANT MATH SERIES — AB & BC — AP CAL LESSON 16

Example 1 Mean Value theorem

If $f(x) = -x^2 + 8x - 6$ on the closed interval $[1, 4]$, find all numbers c that satisfy the conclusion of the Mean Value theorem.

Solution $f(x)$ is a polynomial function. So, f is continuous on $[1, 4]$ and is differentiable on $(1, 4)$. Since $f'(x) = -2x + 8$ and the slope of line that passes through $(1, 1)$ and $(4, 10)$ is $\frac{10-1}{4-1} = 3$,

$$f'(c) = \frac{f(b) - f(a)}{b - a}$$
$$-2c + 8 = 3$$
$$c = \frac{5}{2}$$

Therefore, the number c that satisfy the conclusion of the Mean Value theorem and in the interval $[1, 4]$ is 2.5.

Rolle's Theorem

Let f be a function that satisfies the following three hypotheses:

1. f is continuous on the closed interval $[a, b]$.
2. f is differentiable on the open interval (a, b).
3. $f(a) = f(b)$.

Then there is a c is (a, b) such that $f'(c) = 0$.

Below shows the graphs of functions that satisfy the three hypotheses.

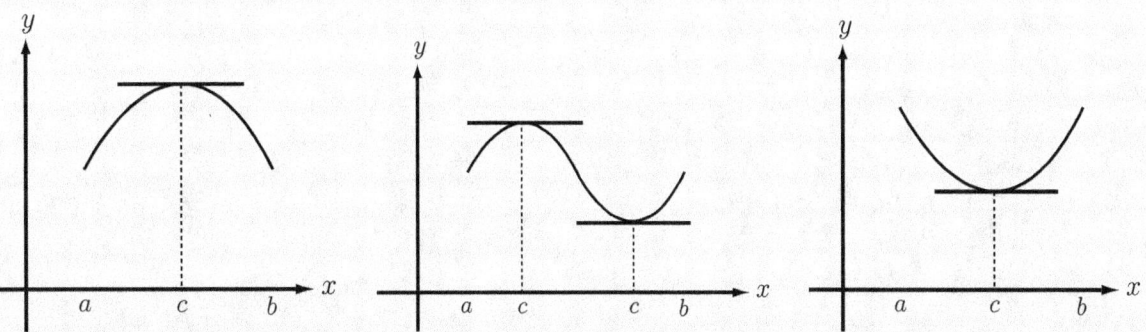

As you can see, there is at least one c in (a, b) such that the tangent line is horizontal, which indicates $f'(c) = 0$.

Tip Rolle's theorem is a special case of the Mean Value theorem where $f(a) = f(b)$.

MR. RHEE'S BRILLIANT MATH SERIES   AB & BC   AP CAL LESSON 16

EXERCISES

For questions 1-4, find all values of c in the given interval that are guaranteed by the Mean Value theorem.

1. $f(x) = 2x^2 + 4x + 5$, $\quad [-1, 2]$

2. $f(x) = \dfrac{x}{x+1}$, $\quad [0, 3]$

3. $f(x) = e^x$, $\quad [0, 2]$

4. $f(x) = x^3 - x^2 - 1$, $\quad [-1, 1]$

MR. RHEE'S BRILLIANT MATH SERIES AB & BC AP CAL LESSON 16

For questions 5-8, find all values of c in the given interval that satisfy the conclusion of the Rolle's theorem.

5. $f(x) = x^2 - 6x + 13$, $[1, 5]$

6. $f(x) = \sin 2x$, $[0, \frac{\pi}{2}]$

7. $f(x) = x^3 - 6x^2 + 11x - 6$, $[1, 2]$

8. $f(x) = x\sqrt{x+3}$, $[-3, 0]$

MR. RHEE'S BRILLIANT MATH SERIES — AB & BC — AP CAL LESSON 16

Answers

1. $c = \dfrac{1}{2}$
2. $c = 1$
3. $c = 1.16$
4. $c = -\dfrac{1}{3}$
5. $c = 3$
6. $c = \dfrac{\pi}{4}$
7. $c = 1.423$
8. $c = -2$

LESSON 17

Understanding A Curve From The First And Second Derivatives

Increasing and Decreasing Test

Increasing and Decreasing Test

1. If $f'(x) > 0$ for all x on an interval, then f is increasing on that interval.
2. If $f'(x) < 0$ for all x on an interval, then f is decreasing on that interval.

Example 1 Finding where a function is increasing or decreasing

Find where the function $f(x) = x^3 - 9x^2 + 24x + 1$ is increasing and where it is decreasing.

Solution $f'(x) = 3x^2 - 18x + 24 = 3(x-2)(x-4)$.

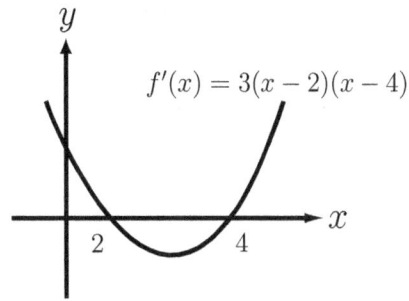

As shown in the figure above, $f'(x) > 0$ for $x < 2$ and also for $x > 4$. Whereas, $f'(x) < 0$ for $2 < x < 4$. Therefore, f is increasing for $x < 2$ and $x > 4$, and f is decreasing for $2 < x < 4$.

Concavity

If the graph of f lies above all of its tangents on an interval, then it is called **concave upward** on that interval. Whereas, if the graph of f lies below all of its tangent on an interval, it is called **concave downward** on that interval.

MR. RHEE'S BRILLIANT MATH SERIES AB & BC AP CAL LESSON 17

Concavity Test

Concavity Test

1. If $f''(x) > 0$ for all x on an interval, then the graph of f is concave upward on that interval.

2. If $f''(x) < 0$ for all x on an interval, then the graph of f is concave downward on that interval.

Tip
1. $f''(x) > 0$ means that $f'(x)$ is increasing. Thus the graph of f is concave upward.
2. $f''(x) < 0$ means that $f'(x)$ is decreasing. Thus the graph of f is concave downward.

Inflection Point

A point P on the curve is called an **inflection point** if the curve changes from concave upward to concave downward or from concave downward to concave upward at P.

If an inflection point on the curve at $x = a$, $f''(a) = 0$.

Example 2 Finding the intervals of concavity and the inflection point

If $f(x) = x^3 - 9x^2 + 24x + 1$, find the intervals of concavity and the inflection point.

Solution $f'(x) = 3x^2 - 18x + 24 = 3(x-2)(x-4)$ and $f''(x) = 6x - 18$. Since $f''(x) = 0$ when $x = 3$, $(3, 19)$ is the inflection point. Using the Concavity test,

$$f''(x) > 0 \quad \text{for } x > 3 \quad \Longrightarrow \quad f \text{ is concave upward}$$
$$f''(x) < 0 \quad \text{for } x < 3 \quad \Longrightarrow \quad f \text{ is concave downward}$$

Therefore, f is concave upward for $x > 3$, and f is concave downward for $x < 3$.

MR. RHEE'S BRILLIANT MATH SERIES — AB & BC — AP CAL LESSON 17

> ### Finding a Local Maximum and a Local Minimum
>
> There are two ways to find a local maximum and a local minimum: **The first derivative test** and **the second derivative test**.
>
> **The First Derivative Test**: Suppose that c is a critical number of a continuous function f.
>
> 1. If f' changes from positive to negative at c, then f has a local maximum at c.
> 2. If f' changes from negative to positive at c, then f has a local minimum at c.
>
> **The Second Derivative Test**: Suppose f'' is continuous near c.
>
> 1. If $f'(c) = 0$ and $f''(c) > 0$, then f has a local minimum at c.
> 2. If $f'(c) = 0$ and $f''(c) < 0$, then f has a local maximum at c.

Example 3   Finding the local maximum and local minimum

If $f(x) = x^3 - 9x^2 + 24x + 1$, find the local maximum and local minimum values of f.

Solution $f'(x) = 3x^2 - 18x + 24 = 3(x-2)(x-4)$ and $f''(x) = 6x - 18$. $f'(x) = 0$ for $x = 2$ and $x = 4$. So, the critical numbers are 2 and 4. Let's find the local maximum and local minimum values of f using both the First Derivative test and Second Derivative test.

- Using the First Derivative Test:

    As shown in the figure below, f' changes positive to negative at $x = 2$. Thus, f has a local maximum at $x = 2$. The local maximum of value of f is $f(2) = 21$.

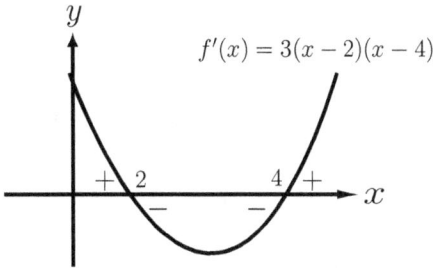

    However, f' changes negative to positive at $x = 4$. Thus, f has a local minimum at $x = 4$. The local minimum of value of f is $f(4) = 17$.

MR. RHEE'S BRILLIANT MATH SERIES AB & BC   AP CAL LESSON 17

- Using the Second Derivative Test:

Since the critical numbers are 2 and 4, substituting these values into the second derivative $f''(x) = 6x - 18$ will determine the local maximum and local minimum value of f.

$$f''(2) = 6(2) - 18 < 0 \quad \implies \quad f \text{ has the local maximum at } x = 2$$
$$f''(4) = 6(4) - 18 > 0 \quad \implies \quad f \text{ has the local minimum at } x = 4$$

Therefore, f has the local maximum of value of f is $f(2) = 21$ and the local minimum of value of f is $f(4) = 17$.

MR. RHEE'S BRILLIANT MATH SERIES AB & BC   AP CAL LESSON 17

EXERCISES

For questions 1-4, find the local maximum and local minimum values of f.

1. $f(x) = x^3 - 3x^2 + 3$

2. $f(x) = x^2 e^x$

3. $f(x) = x - \sqrt{x-1}$

4. $f(x) = \dfrac{x}{x^2 + 1}$

5. If $f''(x) = (x-1)(x-2)^2(x-3)$, find the inflection points.

6. If $f(x) = x \ln x$, find the intervals on which f is increasing.

MR. RHEE'S BRILLIANT MATH SERIES AB & BC   AP CAL LESSON 17

Answers

1. Local maximum is $(0, 3)$,   Local minimum is $(2, -1)$
2. Local maximum is $\left(-2, \dfrac{4}{e^2}\right)$,   Local minimum is $(0, 0)$
3. Local minimum is $\left(\dfrac{5}{4}, \dfrac{3}{4}\right)$
4. Local maximum is $\left(1, \dfrac{1}{2}\right)$,   Local minimum is $\left(-1, -\dfrac{1}{2}\right)$
5. Inflection points are at $x = 1$ and $x = 3$
6. $x > \dfrac{1}{e}$

MR. RHEE'S BRILLIANT MATH SERIES — AB & BC — AP CAL LESSON 18

LESSON 18

Optimization Problems

Optimization Problems

Optimization is one of the most important applications of the first derivative because it has many applications in real life. In general, optimization consists of maximizing or minimizing a function with a constraint. The following guidelines will help you solve optimization problems.

Guidelines for solving maximum and minimum problems

1. Read the problem and draw a diagram.

2. Define variables and label your diagram with these variables. It will help you set up mathematical equations.

3. Set up two equations that are related to the diagram. One equation is an optimization equation and the other is a constraint equation. The constraint equation is used to solve for one of the variables.

4. Set up the optimization equation as a function of only one variable using the constraint equation.

5. Differentiate the optimization equation and find the critical numbers.

6. Find the local maximum and minimum using the first or second derivative tests.

Example 1 Solving optimization problems

Find the two positive numbers whose product is 100 and whose sum is minimum.

Solution Let x and y be the two positive numbers. So, the sum S and constraint can be defined as

$$\text{Minimization equation:} \quad S(x,y) = x + y$$
$$\text{Constraint equation:} \quad xy = 100$$

From the constraint equation, we get $y = \dfrac{100}{x}$. Substituting this into the minimization equation,

$$S(x) = x + \dfrac{100}{x}$$

Differentiate $S(x)$ with respect to x and find the critical numbers.

$$S'(x) = 1 - \frac{100}{x^2} = \frac{x^2 - 100}{x^2}$$

Since x is the positive number, $S'(x) = 0$ when $x = 10$. So, the critical number is 10. Using the First Derivative test, $S'(x) < 0$ when $x < 10$ and $S(x) > 0$ when $x > 10$. Thus, the function $S(x)$ has the local minimum at $x = 10$. Therefore, the sum S is minimum when $x = 10$ and $y = \frac{100}{x} = 10$.

Example 2 Solving optimization problems

A farmer has 1000 m of fencing and wants to fence off a rectangular field that borders a straight river. He does not need any fence along the river. Find the dimensions of the field that has the largest area.

Solution As shown in the figure below, let x and y be the width and length of the rectangular field, respectively.

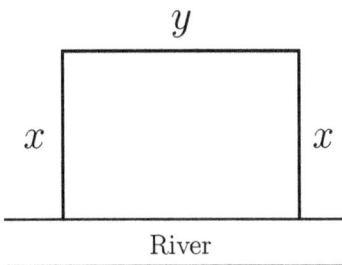

So, the area A and constraint can be defined as

$$\text{Maximization equation:} \quad A(x, y) = xy$$
$$\text{Constraint equation:} \quad 2x + y = 1000$$

From the constraint equation, we get $y = 1000 - 2x$. Substituting this into the maximization equation,

$$A(x) = x(1000 - 2x) = 1000x - 2x^2$$

Differentiate $A(x)$ with respect to x and find the critical numbers.

$$A'(x) = 1000 - 4x$$

Since $A'(x) = 0$ when $x = 250$, the critical number is 250. Note that $A''(x) = -4$ for all x. Using the Second Derivative test, $A''(250) < 0$ which indicates that the function $A(x)$ has the local maximum at $x = 250$. Thus, $y = 1000 - 2x = 1000 - 2(250) = 500$. Therefore, the dimensions of the field that has the largest area are width of 250 m and length of 500 m.

MR. RHEE'S BRILLIANT MATH SERIES   AB & BC   AP CAL LESSON 18

EXERCISES

1. If $2700\,\text{cm}^2$ of material is available to make a box with a square base and an open top, find the largest possible volume of the box.

2. Find the dimensions of the rectangle of largest area that has its base on the x-axis and its other two vertices above the x-axis and lying on the parabola $y = 12 - x^2$.

3. Find the point on the parabola $y^2 = 2x$ that is closed to the point $(1, 4)$.

MR. RHEE'S BRILLIANT MATH SERIES   AB & BC   AP CAL LESSON 18

4. The top and bottom margins of a poster are each 6 inches and the side margins are each 4 inches. If the area of printed material on the poster is fixed at $384\,\text{in}^2$, find the dimensions of the poster with the smallest area.

5. The volume of a rectangular storage container with an open top is $10\,\text{ft}^3$. The length of the its base is twice the width. Material for the base costs \$12 per square feet. Material for the sides costs \$8 per square feet. Find the cost of materials for the cheapest container.

6. Find the area of the largest rectangle that can be inscribed in a semicircle of radius r.

MR. RHEE'S BRILLIANT MATH SERIES — AB & BC — AP CAL LESSON 18

Answers

1. $13500\,cm^3$
2. Length=4, Width=8
3. $(2,2)$
4. Length = 24 in, Width = 36 in
5. $210.53
6. r^2

MR. RHEE'S BRILLIANT MATH SERIES AB & BC AP CAL LESSON 19

LESSON 19

Indefinite Integrals

Antiderivative or Indefinite Integral

The derivatives of four functions below are the same: $3x^2$.

$$(x^3+3)' = 3x^2, \qquad (x^3+2)' = 3x^2, \qquad (x^3+1)' = 3x^2, \qquad (x^3+C)' = 3x^2$$

An antiderivative or indefinite integral of f is a differentiable function F such that the derivative of F is f; that is, $F' = f$. The process of solving for antiderivatives is called **indefinite integration** and its opposite operation is differentiation. For instance, the antiderivative of $3x^2$ is $x^3 + C$, where C is a constant. This can be written as

$$\int 3x^2 dx = x^3 + C$$

In general,

$$\int f dx = F + C \qquad \text{because} \qquad \frac{d}{dx}(F+C) = f$$

Example 1 Finding the antiderivative

Find the antiderivative of $\cos x$.

Solution Since $(\sin x + C)' = \cos x$,

$$\int \cos dx = \sin x + C$$

Thus, the the antiderivative of $\cos x$ is $\sin x + C$.

Example 2 Finding the antiderivative

Find the indefinite integral of $x^3 + 4x + 1$.

Solution Since $\left(\frac{1}{4}x^4 + 2x^2 + x + C\right)' = x^3 + 4x + 1$,

$$\int x^3 + 4x + 1 \, dx = \frac{1}{4}x^4 + 2x^2 + x + C$$

MR. RHEE'S BRILLIANT MATH SERIES
AB & BC — AP CAL LESSON 19

Basic Indefinite Integrals

1. $\int cf(x)dx = c\int f(x)dx$
2. $\int [f(x) \pm g(x)]dx = \int f(x)dx \pm \int g(x)dx$
3. $\int k\,dx = kx + C$
4. $\int x^n = \dfrac{1}{n+1}x^{n+1} + C \quad (n \neq -1)$
5. $\int \dfrac{1}{x} = \ln|x| + C$
6. $\int a^x\,dx = \dfrac{a^x}{\ln a} + C$
7. $\int e^x\,dx = e^x + C$
8. $\int \sin x\,dx = -\cos x + C$
9. $\int \cos x\,dx = \sin x + C$
10. $\int \sec^2 x\,dx = \tan x + C$
11. $\int \csc^2 x\,dx = -\cot x + C$
12. $\int \sec x \tan x\,dx = \sec x + C$
13. $\int \csc x \cot x\,dx = -\csc x + C$
14. $\int \dfrac{1}{x^2+1}\,dx = \tan^{-1} x + C$
15. $\int \dfrac{1}{\sqrt{1-x^2}}\,dx = \sin^{-1} x + C$
16. $\int \tan x\,dx = \ln|\sec x| + C$
17. $\int \cot x\,dx = \ln|\sin x| + C$

 Tip

1. We need to use the Substitution rule to find the indefinite integrals of $\tan x$ and $\cot x$. However, consider these indefinite integrals as basic integrals and memorize them for now.
2. Note that $\int f(x) \cdot g(x)\,dx \neq \int f(x)dx \cdot \int g(x)dx$

Example 3 Finding the indefinite integral

Find the indefinite integral of $f(x) = \sqrt[3]{x^2}$.

Solution Since $\sqrt[3]{x^2}$ can be written as $x^{\frac{2}{3}}$,

$$\int \sqrt[3]{x^2}\,dx = \int x^{\frac{2}{3}}\,dx = \dfrac{1}{\frac{2}{3}+1}x^{\frac{2}{3}+1} + C$$
$$= \dfrac{3}{5}x^{\frac{5}{3}} + C$$

Example 4 Finding the indefinite integral

Find the indefinite integral of $f(x) = \dfrac{x^2 + x + 1}{x}$.

Solution

$$\int \frac{x^2+x+1}{x}dx = \int \frac{x^2}{x} + \frac{x}{x} + \frac{1}{x}dx$$
$$= \int x + 1 + \frac{1}{x}dx$$
$$= \int xdx + \int 1dx + \int \frac{1}{x}dx$$
$$= \frac{1}{2}x^2 + x + \ln|x| + C$$

Example 5 Finding the indefinite integral

Find the indefinite integral of $f(\theta) = \dfrac{\sin 2\theta}{\cos \theta}$.

Solution $\sin 2\theta = 2\sin\theta\cos\theta$.

$$\int \frac{\sin 2\theta}{\cos \theta}d\theta = \int \frac{2\sin\theta\cos\theta}{\cos\theta}d\theta$$
$$= \int 2\sin\theta d\theta$$
$$= 2\int \sin\theta d\theta$$
$$= -2\cos\theta + C$$

Example 6 Finding the indefinite integral

Find the indefinite integral of $f(x) = \dfrac{4}{1+x^2} + 3e^x$.

Solution

$$\int \frac{4}{1+x^2} + 3e^x dx = \int \frac{4}{1+x^2}dx + \int 3e^x dx$$
$$= 4\int \frac{1}{1+x^2}dx + 3\int e^x dx$$
$$= 4\tan^{-1}x + 3e^x + C$$

MR. RHEE'S BRILLIANT MATH SERIES AB & BC AP CAL LESSON 19

EXERCISES

For questions 1-8, Find the indefinite integrals

1. $\int (x-1)(x+2)dx$

2. $\int \frac{1}{x} - \frac{1}{x^2} dx$

3. $\int \frac{\cos^2 x + 1}{\cos^2 x} dx$

4. $\int (x-1)^3 dx$

5. $\int \frac{x+1}{\sqrt{x}} dx$

6. $\int \sec^2 x - \tan^2 x \, dx$

7. $\int \dfrac{\sin x}{1 - \sin^2 x} dx$

8. $\int \dfrac{\cos 2x}{\cos^2 x} dx$

Answers

1. $\frac{1}{3}x^3 + \frac{1}{2}x^2 - 2x + c$
2. $\ln|x| + \frac{1}{x} + C$
3. $x + \tan x + C$
4. $\frac{1}{4}x^4 - x^3 + \frac{3}{2}x^2 - x + C$
5. $\frac{2}{3}x^{\frac{3}{2}} + 2x^{\frac{1}{2}} + C$
6. $x + C$
7. $\sec x + C$
8. $2x - \tan x + C$

MR. RHEE'S BRILLIANT MATH SERIES AB & BC   AP CAL LESSON 20

LESSON 20

The Definite Integral

Definition of a Definite Integral

Suppose f is a continuous function on the closed interval $[a, b]$ as shown in Figure 1. Let's divide the interval $[a, b]$ into n subintervals of equal width $\Delta x = \frac{b-a}{n}$. Choose x_i^* lies in the ith subinterval $[x_{i-1}, x_i]$. Then $f(x_i^*)\Delta x$ represents the area of the ith rectangle.

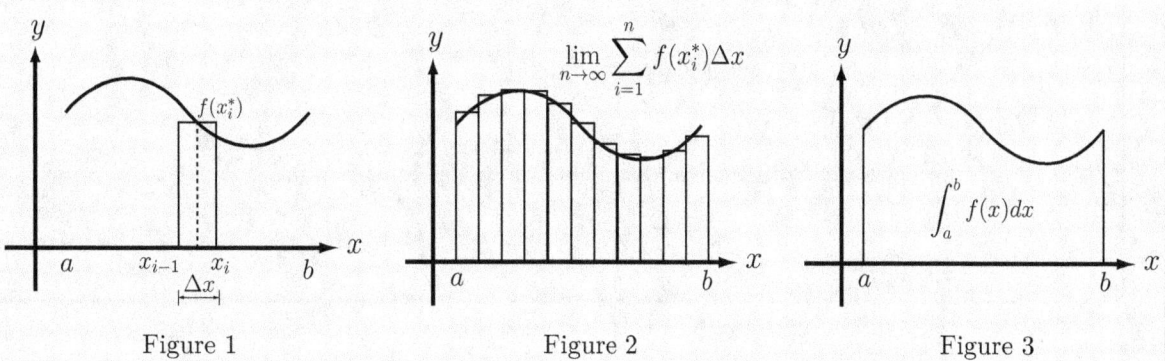

Figure 1   Figure 2   Figure 3

As $n \to \infty$, there are infinitely many rectangles as shown in Figure 2. The sum of areas of infinitely many rectangles can be written as

$$\lim_{n \to \infty} \sum_{i=1}^{n} f(x_i^*)\Delta x$$

is called **Riemann sum**. It can be expressed as

$$\lim_{n \to \infty} \sum_{i=1}^{n} f(x_i^*)\Delta x = \int_a^b f(x)\, dx$$

a definite integral of f from a to b. In the notation of $\int_a^b f(x)\, dx$, the symbol $\int dx$ is called an **integral sign**, $f(x)$ is called the **integrand**, a is called **lower limit** and b is called **upper limit**.

The procedure of calculating an integral is called **integration**. In general, $\int_a^b f(x)\, dx$ represents the area under curve $y = f(x)$ from a to b. as shown in Figure 3.

MR. RHEE'S BRILLIANT MATH SERIES — AB & BC — AP CAL LESSON 20

Express the Riemann Sum as a Definite Integral

Suppose the Riemann sum and $x_i^* = a + i\Delta x$ are given below.

$$\lim_{n\to\infty} \sum_{i=1}^{n} f(x_i^*)\Delta x = \lim_{n\to\infty} \sum_{i=1}^{n} f(a + i\Delta x)\Delta x$$

The following guidelines will help you express the Riemann sum as a definite integral.

Guidelines for expressing the Riemann Sum as a Definite Integral

1. Replace $\lim_{n\to\infty} \sum_{i=1}^{n}$ by $\int$, and Δx by dx.

2. $x_i^* = a + i\Delta x$ indicates that the lower limit is a.

3. Replace $a + i\Delta x$ by x.

4. Since the lower limit a is known, determine the upper limit b from $\Delta x = \frac{b-a}{n}$.

$$\lim_{n\to\infty} \sum_{i=1}^{n} f(a + i\Delta x)\Delta x = \int_{a}^{b} f(x)\, dx$$

Example 1 Expressing the Riemann sum as a Definite Integral

Express $\displaystyle\lim_{n\to\infty} \sum_{i=1}^{n} \left(1 + i\frac{2}{n}\right)^2 \frac{2}{n}$ as a definite integral.

Solution Replace $\displaystyle\lim_{n\to\infty} \sum_{i=1}^{n}$ by $\int$, and $\Delta x = \frac{2}{n}$ by dx as shown below.

$$\lim_{n\to\infty} \sum_{i=1}^{n} \left(1 + i\frac{2}{n}\right)^2 \frac{2}{n} \implies \int \left(1 + i\frac{2}{n}\right)^2 dx$$

Since $x_i^* = a + i\Delta x = 1 + i\frac{2}{n}$ which indicates that the lower limit is $a = 1$. Replace $x_i^* = 1 + i\frac{2}{n}$ by x as shown below.

$$\lim_{n\to\infty} \sum_{i=1}^{n} \left(1 + i\frac{2}{n}\right)^2 \frac{2}{n} \implies \int_{1} (x)^2 \, dx$$

Let's determine the upper limit b. Since $a = 1$ and $\Delta x = \frac{b-a}{n} = \frac{b-1}{n} = \frac{2}{n}$, the upper limit b is $b = 3$. Thus, $\displaystyle\lim_{n\to\infty} \sum_{i=1}^{n} \left(1 + i\frac{2}{n}\right)^2 \frac{2}{n}$ can be written as

$$\lim_{n\to\infty} \sum_{i=1}^{n} \left(1 + i\frac{2}{n}\right)^2 \frac{2}{n} = \int_{1}^{3} x^2 \, dx$$

MR. RHEE'S BRILLIANT MATH SERIES AB & BC AP CAL LESSON 20

Properties of the Definite Integral

1. $\int_a^b f(x)\,dx = -\int_b^a f(x)\,dx$

2. $\int_a^a f(x)\,dx = 0$

3. $\int_a^b c\,dx = c(b-a)$

4. $\int_a^b f(x) \pm g(x)\,dx = \int_a^b f(x)\,dx \pm \int_a^b g(x)\,dx$

5. $\int_a^b cf(x)\,dx = c\int_a^b f(x)\,dx$

6. $\int_a^b f(x)\,dx + \int_b^c f(x)\,dx = \int_a^c f(x)\,dx$

7. If $f(x)\,dx \geq 0$ for $a \leq x \leq b$, then $\int_a^b f(x)\,dx \geq 0$.

8. If $g(x) \leq g(x)$ for $a \leq x \leq b$, then $\int_a^b g(x)\,dx \leq \int_a^b f(x)\,dx$.

Tip Note that the area is positive if the function lies above the x-axis. Whereas, the area is negative if the function lies below the x-axis. Thus, the A_1 is positive and A_2 is negative as shown in the figure below.

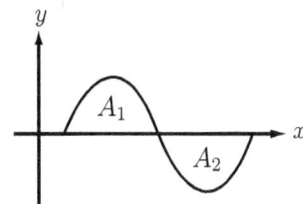

Example 2 Evaluating the areas

Evaluate $\int_0^3 x - 2\,dx$.

Solution $\int_0^3 x - 2\,dx$ represents the area under a line $y = x - 2$ from $x = 0$ and $x = 3$, which is the sum of the areas of the two triangles, A_1 and A_2, as shown below. Since A_1 is $\frac{1}{2}$ and A_2 is -2,

$$\int_0^3 x - 2\,dx = A_1 + A_2 = \frac{1}{2} - 2 = -\frac{3}{2}.$$

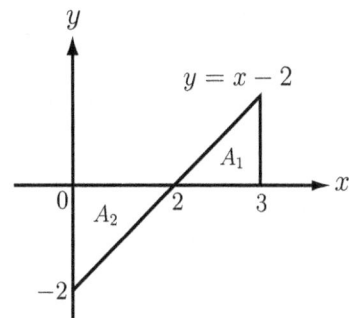

MR. RHEE'S BRILLIANT MATH SERIES		AB & BC		AP CAL LESSON 20

EXERCISES

For questions 1-3, express the following Riemann sums as definite integrals.

1. $\displaystyle\lim_{n\to\infty} \sum_{i=1}^{n} 3\left(2 + i\frac{3}{n}\right)\frac{3}{n}$

2. $\displaystyle\lim_{n\to\infty} \sum_{i=1}^{n} \left(1 + \frac{2i}{n}\right)^3 \frac{2}{n}$

3. $\displaystyle\lim_{n\to\infty} \frac{\pi}{n} \sum_{i=1}^{n} \sin\left(\frac{2\pi i}{n}\right)$

For questions 4-5, express the following definite integrals as the Riemann sums.

4. $\displaystyle\int_1^2 \ln x \, dx$

5. $\displaystyle\int_{\frac{\pi}{2}}^{\pi} \cos x \, dx$

MR. RHEE'S BRILLIANT MATH SERIES	AB & BC	AP CAL LESSON 20

6. If $\int_{-1}^{4} f(x)\,dx = 7$ and $\int_{3}^{4} f(x)\,dx = 3$, evaluate $\int_{-1}^{3} f(x)\,dx$.

For questions 7-10, evaluate the following definite integrals by interpreting it as the area under the curve.

7. $\int_{0}^{\pi} \cos x\,dx$

8. $\int_{0}^{6} |x - 4|\,dx$

9. $\int_{1}^{3} 2\,dx$

10. $\int_{-2}^{2} \sqrt{4 - x^2}\,dx$

MR. RHEE'S BRILLIANT MATH SERIES — AB & BC — AP CAL LESSON 20

Answers

1. $\displaystyle\int_2^5 3x\,dx$
2. $\displaystyle\int_1^3 x^3\,dx$
3. $\displaystyle\int_0^\pi \sin 2x\,dx$
4. $\displaystyle\lim_{n\to\infty}\sum_{i=1}^n \ln\left(1+i\frac{1}{n}\right)\frac{1}{n}$
5. $\displaystyle\lim_{n\to\infty}\sum_{i=1}^n \cos\left(\frac{\pi}{2}+i\frac{\pi}{2n}\right)\frac{\pi}{2n}$
6. 4
7. 0
8. 10
9. 4
10. 2π

MR. RHEE'S BRILLIANT MATH SERIES AB & BC   AP CAL LESSON 21

LESSON 21

Numerical Approximations Of Integration

There are four types of approximations of a definite integral: left endpoint approximation, right endpoint approximation, midpoint rule, and trapezoidal rule.

Left Endpoint Approximation and Right Endpoint Approximation

Suppose we divide $[a, b]$ into n subintervals of equal length $\Delta x = \frac{b-a}{n}$. Then,

$$\int_a^b f(x)\,dx \approx \sum_{i=1}^n f(x_i^*)\Delta x$$

where x_i^* is any point in the ith subinterval $[x_{i-1}, x_i]$. If x_{i-1} is chosen to be the left endpoint of the interval as shown in Figure 1, then $x_i^* = x_{i-1}$. Thus,

$$\int_a^b f(x)\,dx \approx L_n = \sum_{i=1}^n f(x_{i-1})\Delta x$$

L_n is called the **Left endpoint approximation**.

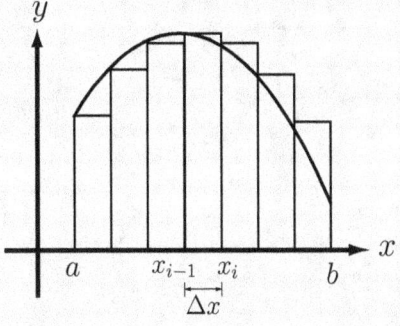
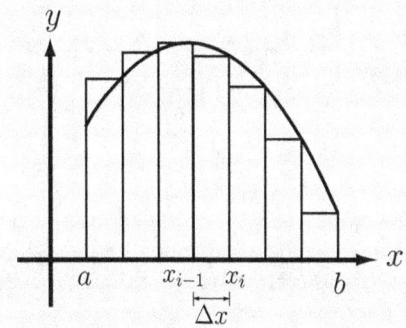

Figure 1   Figure 2

Whereas, if x_i is chosen to be the right endpoint of the interval as shown in Figure 2, then $x_i^* = x_i$. Thus,

$$\int_a^b f(x)\,dx \approx R_n = \sum_{i=1}^n f(x_i)\Delta x$$

R_n is called the **Right endpoint approximation**.

(Tip) 1. If f is increasing on $[a, b]$, $\quad L_n < \int_a^b f(x)\,dx < R_n$

2. If f is decreasing on $[a, b]$, $\quad R_n < \int_a^b f(x)\,dx < L_n$

Midpoint Approximation and Trapezoidal Approximation

Suppose we divide $[a, b]$ into n subintervals of equal length $\Delta x = \frac{b-a}{n}$. Then,

$$\int_a^b f(x)\, dx \approx \sum_{i=1}^n f(x_i^*)\Delta x$$

where x_i^* is any point in the ith subinterval $[x_{i-1}, x_i]$. If $\frac{x_{i-1}+x_i}{2}$ is chosen to be the midpoint of the interval as shown in Figure 3, then $x_i^* = \frac{x_{i-1}+x_i}{2}$. Thus,

$$\int_a^b f(x)\, dx \approx M_n = \sum_{i=1}^n f\left(\frac{x_{i-1}+x_i}{2}\right)\Delta x$$

M_n is called the **Midpoint Rule** or **Midpoint approximation**.

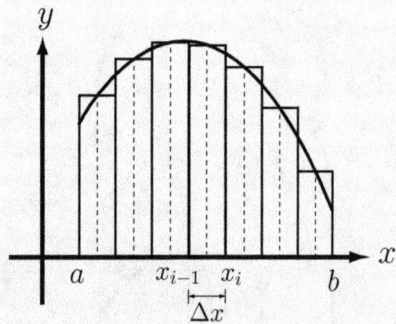

Figure 3

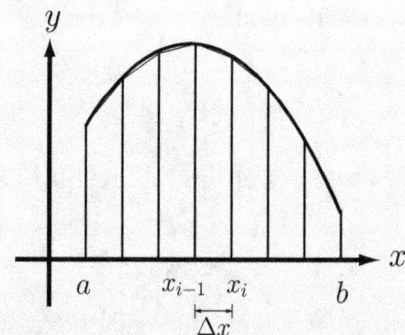

Figure 4

In general, the sum of areas of trapezoids gives you a better approximation of a definite integral than the sum of areas of rectangles over n subintervals. As shown in Figure 4, the area of trapezoid that lies above the ith subinterval can be written as

$$\frac{1}{2}\Big(f(x_{i-1}) + f(x_i)\Big)\Delta x$$

Thus,

$$\int_a^b f(x)\, dx \approx T_n = \sum_{i=1}^n \frac{1}{2}\Big(f(x_{i-1}) + f(x_i)\Big)\Delta x$$

$$= \frac{\Delta x}{2}[f(x_0) + 2f(x_1) + 2f(x_2) + \cdots + 2f(x_{n-1}) + f(x_n)]$$

where $x_0 = a$ and $x_n = b$. T_n is called **Trapezoidal Rule** or **Trapezoidal approximation**.

> **Tip**
>
> 1. If f is concave up on $[a, b]$, $\quad M_n < \int_a^b f(x)\, dx < T_n$
>
> 2. If f is concave down on $[a, b]$, $\quad T_n < \int_a^b f(x)\, dx < M_n$

MR. RHEE'S BRILLIANT MATH SERIES AB & BC AP CAL LESSON 21

Example 1 Approximating a definite integral

Use the left endpoint approximation with $n = 4$ to approximate the integral $\int_1^7 x^2 \, dx$.

Solution Interval width Δx is $\Delta x = \frac{b-a}{n} = \frac{7-1}{4} = 1.5$ and the left endpoints are 1, 2.5, 4, and 5.5.

$$\int_1^7 x^2 \, dx \approx L_4 = \sum_{i=1}^{4} f(x_{i-1})\Delta x$$
$$= f(1)\Delta x + f(2.5)\Delta x + f(4)\Delta x + f(5.5)\Delta x$$
$$= \Delta x \Big(f(1) + f(2.5) + f(4) + f(5.5) \Big)$$
$$= 1.5(1 + 6.25 + 16 + 30.25)$$
$$= 80.25$$

Example 2 Approximating a definite integral

Use the Trapezoidal Rule with $n = 5$ to approximate the integral $\int_1^3 \frac{1}{x} \, dx$.

Solution Interval width Δx is $\Delta x = \frac{b-a}{n} = \frac{3-1}{5} = 0.4$. So, the Trapezoidal Rule gives

$$\int_1^3 \frac{1}{x} \, dx \approx T_5 = \sum_{i=1}^{5} \frac{1}{2}\Big(f(x_{i-1}) + f(x_i)\Big)\Delta x$$
$$= \frac{\Delta x}{2}[f(x_0) + 2f(x_1) + 2f(x_2) + \cdots + 2f(x_{n-1}) + f(x_n)]$$
$$= \frac{0.4}{2}[f(1) + 2f(1.4) + 2f(1.8) + 2f(2.2) + 2f(2.6) + f(3)]$$
$$= 0.2\Big(\frac{1}{1} + \frac{2}{1.4} + \frac{2}{1.8} + \frac{2}{2.2} + \frac{2}{2.6} + \frac{1}{3}\Big)$$
$$\approx 1.1103$$

EXERCISES

1. Use the right endpoint with $n = 4$ to approximate the integral $\int_0^4 \sqrt{x}\, dx$.

2. Use the left endpoint with $n = 5$ to approximate the integral $\int_0^1 e^x\, dx$.

3. Use the Midpoint Rule with $n = 4$ to approximate the integral $\int_0^2 x^2 + 1\, dx$.

x	2	4	6	8	10
$f(x)$	10	13	9	11	14

4. Use a Trapezoidal rule with the four subintervals of equal length indicated by the table shown above to approximate $\int_2^{10} f(x)\, dx$.

5. Given the definite integral $\int_1^2 \ln x \, dx$,

   (a) Use a Trapezoidal rule with the five subintervals approximate the definite integral.

   (b) Is your answer to part (a) an overestimate or an underestimate? Justify your answer.

t	0	3	4	6	10
$v(t)$	2	5	7	4	6

6. The table above shows the velocity of a runner during the first 10 seconds. Use the Trapezoidal Rule with four subintervals of unequal length to approximate the distance covered by the runner during those 10 seconds.

MR. RHEE'S BRILLIANT MATH SERIES — AB & BC — AP CAL LESSON 21

Answers

1.	6.146
2.	1.552
3.	4.625
4.	90
5(a).	0.385
5(b).	$T_5 = 0.385$, an underestimate
6.	47.5

MR. RHEE'S BRILLIANT MATH SERIES   AB & BC   AP CAL LESSON 22

LESSON 22

The Fundamental Theorem Of Calculus

The Fundamental Theorem of Calculus establishes a connection between the differentiation and integration such that differentiation and integration are inverse processes.

The Fundamental Theorem of Calculus, Part I

If f is continuous on $[a, b]$, then the function g defined by

$$g(x) = \int_a^x f(t)\, dt \qquad a \leq x \leq b$$

is continuous on $[a, b]$ and differentiable on (a, b), and

$$g'(x) = \frac{d}{dx} \int_a^x f(t)\, dt \quad \Longrightarrow \quad g'(x) = f(x)$$

Tip

1. The Fundamental Theorem of Calculus, Part I is important because it guarantees that the existence of antiderivatives for continuous functions.

2. You can choose any arbitrary number for the lower limit a because $g'(x)$ is not affected by a. For instance,

$$g(x) = \int_2^x f(t)\, dt, \quad \Longrightarrow \quad g'(x) = \frac{d}{dx}\int_2^x f(t)\, dt = f(x)$$

$$g(x) = \int_0^x f(t)\, dt, \quad \Longrightarrow \quad g'(x) = \frac{d}{dx}\int_0^x f(t)\, dt = f(x)$$

3. In case the upper limit is $p(x)$, then $g'(x)$ is

$$g'(x) = \frac{d}{dx}\int_a^{p(x)} f(t)\, dt \quad \Longrightarrow \quad g'(x) = f(p(x)) \cdot (p(x))'$$

For instance, if $g(x) = \int_0^{x^2} \sin t\, dt$, then $g'(x)$ is

$$g'(x) = \frac{d}{dx}\int_0^{x^2} \sin t\, dt \quad \Longrightarrow \quad g'(x) = \sin(x^2) \cdot (2x)$$

4. In case the lower limit is $q(x)$ and upper limit is $p(x)$, then $g'(x)$ is

$$g'(x) = \frac{d}{dx}\int_{q(x)}^{p(x)} f(t)\, dt \quad \Longrightarrow \quad g'(x) = f(p(x)) \cdot (p(x))' - f(q(x)) \cdot (q(x))'$$

MR. RHEE'S BRILLIANT MATH SERIES AB & BC AP CAL LESSON 22

> ### The Fundamental Theorem of Calculus, Part II
>
> If f is continuous on $[a, b]$, then
>
> $$\int_a^b f(x)\, dx = F(x) \Big]_a^b = F(b) - F(a)$$
>
> Where F is any antiderivative of f such that $F' = f$.

Tip The Fundamental Theorem of Calculus, Part II enables us to simplify the computation of definite integrals using any antiderivative of f.

Example 1 Finding the derivative using the Fundamental Theorem of Calculus, Part I

If $g(x) = \displaystyle\int_1^{2x} \sqrt{3t+1}\, dt$, find $g'(x)$.

Solution

$$g'(x) = \frac{d}{dx} \int_1^{2x} \sqrt{3t+1}\, dt = \sqrt{3(2x)+1} \cdot (2x)' = 2\sqrt{6x+1}$$

Example 2 Finding the derivative using the Fundamental Theorem of Calculus, Part I

If $g(x) = \displaystyle\int_{\sin x}^{\tan x} t\, dt$, find $g'(x)$.

Solution

$$g'(x) = \frac{d}{dx} \int_{\sin x}^{\tan x} t\, dt = \tan x \cdot (\tan x)' - \sin x \cdot (\sin x)'$$
$$= \tan x \sec^2 x - \sin x \cos x$$

MR. RHEE'S BRILLIANT MATH SERIES — AB & BC — AP CAL LESSON 22

Example 3 Evaluate the integral using the Fundamental Theorem of Calculus, Part 2

Evaluate $\int_2^1 \dfrac{1}{x^2}\,dx$.

Solution Since the antiderivative of $\dfrac{1}{x^2}$ is $-\dfrac{1}{x}$,

$$\int_2^1 \dfrac{1}{x^2}\,dx = -\int_1^2 \dfrac{1}{x^2}\,dx$$

$$= \dfrac{1}{x}\bigg]_1^2 = \left(\dfrac{1}{2} - \dfrac{1}{1}\right) = -\dfrac{1}{2}$$

The Total Change Theorem

Let $F'(x)$ be a rate of change. Then, the integral of a rate of change given by $\int_a^b F'(x)\,dx$ is

$$\int_a^b F'(x)\,dx = F(x)\bigg]_a^b = F(b) - F(a)$$

and is the total change such that $F(b) - F(a)$.

Tip

1. If $F'(x)$ is the rate of growth of a population,

$$\int_{t_1}^{t_2} F'(x)\,dx = F(t_2) - F(t_1)$$

is the total change in population from time period $t = t_1$ to $t = t_2$.

2. If $F'(x)$ is the velocity of an object,

$$\int_{t_1}^{t_2} F'(x)\,dx = F(t_2) - F(t_1)$$

is the total change in position, or **displacement** of the particle from time period $t = t_1$ to $t = t_2$.

Difference between Displacement and Total Distance

If an object moves along a straight line with position function $s(t)$, then its velocity is $v(t) = s'(t)$. Displacement and total distance are given by

$$\text{Displacement} = \int_{t_1}^{t_2} v(t)\,dt, \qquad \text{Total Distance} = \int_{t_1}^{t_2} |v(t)|\,dt$$

From the figure below, notice that A_2 is negative since the velocity function $v(t)$ lies below the x-axis. Thus, both displacement and total distance can be interpreted in terms of areas under a velocity curve.

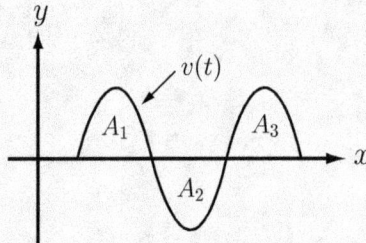

$$\text{Displacement} = \int_{t_1}^{t_2} v(t)\,dt = A_1 + A_2 + A_3, \qquad \text{Total Distance} = \int_{t_1}^{t_2} |v(t)|\,dt = A_1 - A_2 + A_3$$

MR. RHEE'S BRILLIANT MATH SERIES AB & BC   AP CAL LESSON 22

EXERCISES

1. If $g(x) = \int_1^{x^2} \sin t \, dt$, find $g'(x)$.

2. If $g(x) = \int_{2x}^{-1} e^t \, dt$, find $g'(x)$.

3. Evaluate $\int_2^3 x^2 \, dx$.

4. Evaluate $\int_1^4 \dfrac{1}{\sqrt{x}} + 1 \, dx$.

MR. RHEE'S BRILLIANT MATH SERIES   AB & BC   AP CAL LESSON 22

5. Evaluate $\displaystyle\int_1^{\sqrt{3}} \frac{1}{1+x^2}\, dx$.

6. Evaluate $\displaystyle\int_{-1}^{1} \frac{1}{x}\, dx$.

7. Evaluate $\displaystyle\int_{\frac{\pi}{2}}^{\frac{3\pi}{2}} \sin x\, dx$.

8. Suppose a particle moves along a straight line with its velocity at time t is given by $v(t) = t^2 - 3t + 2$, where $0 \leq t \leq 4$ (measured in meters per second).

   (a) Find the displacement of the particle during the time period.

   (b) Find the total distance traveled by the particle during the time period.

MR. RHEE'S BRILLIANT MATH SERIES   AB & BC   AP CAL LESSON 22

Answers

1. $g'(x) = 2x \sin x^2$
2. $g'(x) = -2e^{2x}$
3. $\dfrac{19}{3}$
4. 5
5. $\dfrac{\pi}{12}$
6. Does not exist
7. 0
8 (a). 5.333
8 (b). 5.667

MR. RHEE'S BRILLIANT MATH SERIES AB & BC AP CAL LESSON 23

LESSON 23

The U-Substitution Rule

Basic Indefinite Integrals

Until now, we have evaluated many indefinite integrals using the basic indefinite integrals shown below.

1. $\int cf(x)dx = c\int f(x)dx$
2. $\int [f(x) \pm g(x)]dx = \int f(x)dx \pm \int g(x)dx$
3. $\int k\,dx = kx + C$
4. $\int x^n = \dfrac{1}{n+1}x^{n+1} + C \quad (n \neq -1)$
5. $\int \dfrac{1}{x} = \ln|x| + C$
6. $\int a^x dx = \dfrac{a^x}{\ln a} + C$
7. $\int e^x dx = e^x + C$
8. $\int \sin x\,dx = -\cos x + C$
9. $\int \cos x\,dx = \sin x + C$
10. $\int \sec^2 x\,dx = \tan x + C$
11. $\int \csc^2 x\,dx = -\cot x + C$
12. $\int \sec x \tan x\,dx = \sec x + C$
13. $\int \csc x \cot x\,dx = -\csc x + C$
14. $\int \dfrac{1}{x^2+1}dx = \tan^{-1} x + C$
15. $\int \dfrac{1}{\sqrt{1-x^2}}dx = \sin^{-1} x + C$
16. $\int \tan x\,dx = \ln|\sec x| + C$
17. $\int \cot x\,dx = \ln|\sin x| + C$

However, using the basic indefinite integrals, we are not able to evaluate an integral such as

$$\int 2x \cos(x^2)\,dx$$

In order to find this integral, we use a method called **U-Substitution Rule**, which is a valuable tool to find the antiderivative of functions resulted from the Chain rule.

Tip Notice that $\sin(x^2)$ is an antiderivative of $2x\cos(x^2)$ because

$$\left(\sin x^2\right)' = \cos\left(x^2\right) \cdot (x^2)' = 2x\cos(x^2)$$

Thus,

$$\int 2x\cos(x^2)\,dx = \sin(x^2) + C$$

| MR. RHEE'S BRILLIANT MATH SERIES | AB & BC | AP CAL LESSON 23 |

U-Substitution Rule

Suppose f is continuous on $[a,b]$ and $u = g(x)$ is a differentiable function on $[a,b]$. Then, $du = g'(x)dx$. Thus,

$$\int f\big(g(x)\big)g'(x)\,dx = \int f(u)\,du$$

Tip

1. The most important part of the U-Substitution Rule is to change from the variable x to a new variable u, and change from dx to du. The new integral $\int f(u)\,du$ after the U-Substitution becomes one of the basic indefinite integrals so that you can evaluate the new integral at ease.

2. du is the differential. For instance, if $u = x^2$, then $du = 2x\,dx$.

Example 1 Evaluating an indefinite integral using the U-Substitution rule

Evaluate $\int 2x \cos(x^2)\,dx$

Solution Let $u = x^2$. Then $du = 2x\,dx$. Thus,

$$\begin{aligned}
\int 2x \cos(x^2)\,dx &= \int \cos(x^2) 2x\,dx \\
&= \int \cos u\,du &&\int \cos u\,du \text{ is a basic indefinite integral} \\
&= \sin u + C &&\text{Substitute } u \text{ for } x^2 \\
&= \sin x^2 + C
\end{aligned}$$

MR. RHEE'S BRILLIANT MATH SERIES — AB & BC — AP CAL LESSON 23

Example 2 Evaluating an indefinite integral using the U-Substitution rule

Evaluate $\int \dfrac{\ln x}{x}\, dx$.

Solution Let $u = \ln x$. Then $du = \dfrac{1}{x}\, dx$. Thus,

$$\int \frac{\ln x}{x}\, dx = \int \ln x \, \frac{1}{x}\, dx$$

$$= \int u\, du \qquad \qquad \int u\, du \text{ is a basic indefinite integral}$$

$$= \frac{1}{2} u^2 + C \qquad \qquad \text{Substitute } u \text{ for } \ln x$$

$$= \frac{1}{2} \ln^2 x + C$$

U-Substitution for Definite Integrals

Suppose $u = g(x)$ and $du = g'(x)dx$. Then the new lower limit and the upper limit for the integration are $g(a)$ and $g(b)$, respectively. Thus, the definite integral by U-Substitution rule is given by

$$\int_a^b f\big(g(x)\big) g'(x)\, dx = \int_{g(a)}^{g(b)} f(u)\, du$$

Example 3 Evaluating an definite integral using the U-Substitution rule

Evaluate $\int_1^5 \sqrt{x-1}\, dx$.

Solution Let $u = x - 1$. Then $du = dx$. Let's find the new lower limit and upper limit for the integration.

When $x = 1$, $u = x - 1 = 0$, \qquad When $x = 5$, $u = x - 1 = 4$

Thus, the new lower limit and upper limit for the integration are 0 and 4, respectively.

$$\int_1^5 \sqrt{x-1}\, dx = \int_0^4 \sqrt{u}\, du = \frac{2}{3} u^{\frac{3}{2}} \Big]_0^4 = \frac{16}{3}$$

MR. RHEE'S BRILLIANT MATH SERIES   AB & BC   AP CAL LESSON 23

Example 4   **Evaluating an definite integral using the U-Substitution rule**

Evaluate $\displaystyle\int_0^1 \frac{1}{2x+5}\,dx$.

Solution   Let $u = 2x+5$. Then $du = 2dx$, which gives $\frac{1}{2}du = dx$. Let's find the new lower limit and upper limit for the integration.

$$\text{When } x = 0,\ u = 2x+5 = 5, \qquad \text{When } x = 1,\ u = 2x+5 = 7$$

Thus, the new lower limit and upper limit for the integration are 5 and 7, respectively.

$$\begin{aligned}\int_0^1 \frac{1}{2x+5}\,dx &= \int_5^7 \frac{1}{u}\frac{1}{2}\,du = \frac{1}{2}\ln|u|\Big]_5^7 \\ &= \frac{1}{2}(\ln 7 - \ln 5) \\ &= \frac{1}{2}\ln\left(\frac{7}{5}\right)\end{aligned}$$

MR. RHEE'S BRILLIANT MATH SERIES — AB & BC — AP CAL LESSON 23

EXERCISES

1. Evaluate $\int \tan x \, dx$.

2. Evaluate $\int \dfrac{1}{3 - 2x} \, dx$.

3. Evaluate $\int \dfrac{\tan^{-1} x}{1 + x^2} \, dx$.

4. Evaluate $\int \sec x \tan x \sqrt{2 + \sec x} \, dx$.

5. Evaluate $\int \dfrac{1}{x \ln x} \, dx$.

6. Evaluate $\displaystyle\int \frac{x+1}{x^2+1}\,dx$.

7. Evaluate $\displaystyle\int \frac{x}{1+x^4}\,dx$.

8. Evaluate $\displaystyle\int_0^1 xe^{-x^2}\,dx$.

9. Evaluate $\displaystyle\int_1^2 x^2\sqrt{x-1}\,dx$.

10. Evaluate $\displaystyle\int_0^{\frac{\sqrt{2}}{2}} \frac{\sin^{-1} x}{\sqrt{1-x^2}}\,dx$.

MR. RHEE'S BRILLIANT MATH SERIES — AB & BC — AP CAL LESSON 23

Answers

1. $\ln|\sec x| + C$
2. $-\dfrac{1}{2}\ln|3 - 2x| + C$
3. $\dfrac{1}{2}(\tan^{-1} x)^2 + C$
4. $\dfrac{2}{3}(2 + \sec x)^{\frac{3}{2}} + C$
5. $\ln|\ln x| + C$
6. $\dfrac{1}{2}\ln|x^2 + 1| + \tan^{-1} x + C$
7. $\dfrac{1}{2}\tan^{-1}(x^2) + C$
8. $\dfrac{1}{2}\left[1 - \dfrac{1}{e}\right]$
9. $\dfrac{184}{105}$
10. $\dfrac{\pi^2}{32}$

LESSON 24
Area Between Curves

Area Between Curves

Consider the region S enclosed by two curves f and g, and two vertical lines $x = a$ and $x = b$, where f and g are continuous functions in $[a, b]$ and $f(x) \geq g(x)$ for all x in $[a, b]$ as shown in Figure 1. The area of the ith rectangle with width Δx and height $f(x_i^*) - g(x_i^*)$ is $[f(x_i^*) - g(x_i^*)]\Delta x$.

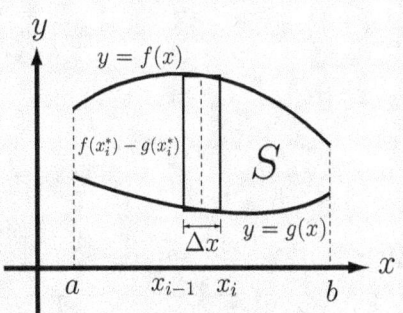

Figure 1

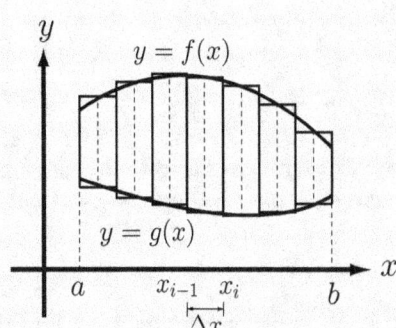

Figure 2

and the area of S can be approximated by the sum of areas of many rectangles. As $n \to \infty$, there are infinitely many rectangles as shown in Figure 2. The sum of areas of infinitely many rectangles can be written as

$$\text{Area of } S = \lim_{n \to \infty} \sum_{i=1}^{n} [f(x_i^*) - g(x_i^*)]\Delta x$$

and it can be expressed

$$\text{Area of } S = \lim_{n \to \infty} \sum_{i=1}^{n} [f(x_i^*) - g(x_i^*)]\Delta x = \int_a^b [f(x) - g(x)]\, dx$$

as a definite integral as shown in Figure 3.

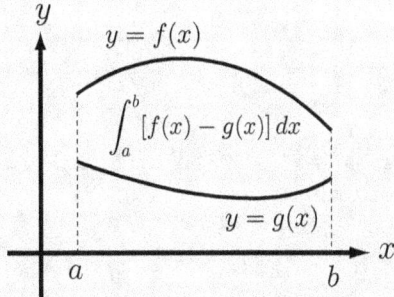

MR. RHEE'S BRILLIANT MATH SERIES — AB & BC — AP CAL LESSON 24

Finding Area between Curves Using ith Vertical Rectangle

Suppose $y = f_T$ and $y = f_B$ are continuous functions in $[a, b]$ and $f_T(x) \geq f_B(x)$ for all x in $[a, b]$ as shown below. f_T and f_B represent the top curve and bottom curve, respectively.

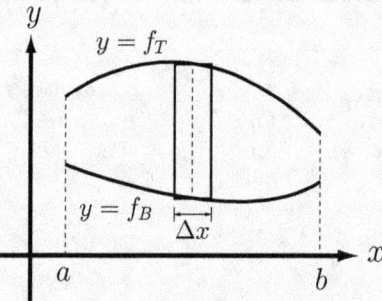

The area of ith vertical rectangle is $(f_T - f_B)\Delta x$. Thus, the area A enclosed by region enclosed by two curves f_T and f_B, and two vertical lines $x = a$ and $x = b$ is

$$A = \int_a^b [f_T - f_B]\, dx$$

Finding Area between Curves Using ith Horizontal Rectangle

Suppose $x = f_R$ and $x = f_L$ are continuous functions and $f_R \geq f_L$ for $c \leq y \leq d$ as shown below. f_R and f_L represent the right curve and left curve, respectively.

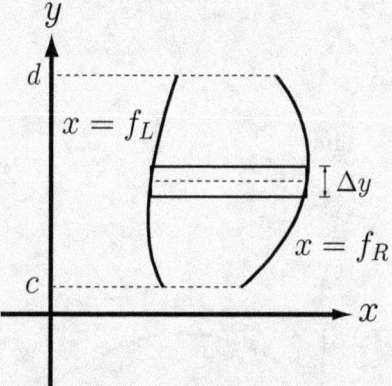

The area of ith horizontal rectangle is $(f_R - f_L)\Delta y$. Thus, the area A enclosed by region enclosed by two curves f_R and f_L, and two horizontal lines $y = c$ and $y = d$ is

$$A = \int_c^d [f_R - f_L]\, dy$$

Example 1 Finding the area between curves

Find the area of the region enclosed by $y = x^2 - 3x$ and $y = 2x$.

Solution Sketch the graphs of $y = 2x$ and $y = x^2 - 3x$ as shown below.

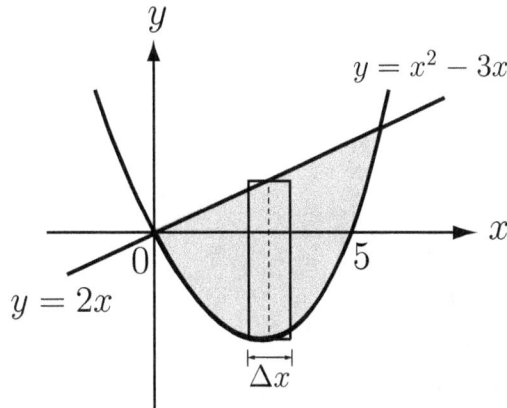

Set $x^2 - 3x = 2x$ and and solve for x to find the intersection points. This gives $x^2 - 5x = 0$ or $x(x - 5) = 0$. Thus, $x = 0$ or $x = 5$. Draw ith rectangle to determine the top curve f_T and the bottom curve f_B. So, $f_T = 2x$ and $f_B = x^2 - 3x$. The area of ith rectangle is $(f_T - f_B)\Delta x$ and the total area of region enclosed by $y = x^2 - 3x$ and $y = 2x$ from $x = 0$ and $x = 5$ is

$$\begin{aligned}
A &= \int_a^b [f_T - f_B]\, dx \\
&= \int_0^5 [2x - (x^2 - 3x)]\, dx \\
&= \int_0^5 (-x^2 + 5x)\, dx \\
&= -\frac{1}{3}x^3 + \frac{5}{2}x^2 \bigg]_0^5 \\
&= -\frac{1}{3}5^3 + \frac{5}{2}5^2 \\
&= \frac{125}{6}
\end{aligned}$$

MR. RHEE'S BRILLIANT MATH SERIES AB & BC AP CAL LESSON 24

EXERCISES

For questions 1-8, sketch the region enclosed by the following curves and find the area of region.

1. $y = x^2 + 1$, $y = x - 1$, $x = -1$, $x = 3$

2. $y = \cos x$, $y = \sin x$, $x = \dfrac{\pi}{4}$, $x = \dfrac{\pi}{2}$

3. $y = x^2$, $y = \sqrt{x}$

4. $x = y^2 + 1$, $x = -y^2 - 1$, $y = -1$, $y = 2$

5. $y = \dfrac{1}{x}$, $y = \dfrac{1}{x^2}$, $x = \dfrac{1}{2}$, $x = 2$

6. $y = e^x$, $y = \sin 2x$, $x = 0$, $x = \dfrac{\pi}{2}$

7. $y = (x-3)^2$, $y = x - 3$

8. $x = \dfrac{1}{2}(y^2 - 6)$, $y = x - 1$

MR. RHEE'S BRILLIANT MATH SERIES — AB & BC — AP CAL LESSON 24

Answers

1. $\dfrac{40}{3}$
2. $\sqrt{2}-1$
3. $\dfrac{1}{3}$
4. 12
5. $\dfrac{1}{2}$
6. $e^{\frac{\pi}{2}}-2$
7. $\dfrac{1}{6}$
8. 18

LESSON 25

Average Value Of A Function and Arc Length

Average Value of a Function

If f is continuous on $[a,b]$, then there exists a number called **average value of** f in $[a,b]$ such that

$$f_{ave} = \frac{1}{b-a} \int_a^b f(x)\, dx$$

In Figure 1, $\int_a^b f(x)\, dx$ represents the area under curve. In Figure 2, there is a rectangle whose length is $(b-a)$. Finding the average value of f is to find the width of the rectangle f_{ave} such that the area of the rectangle equals the area under curve.

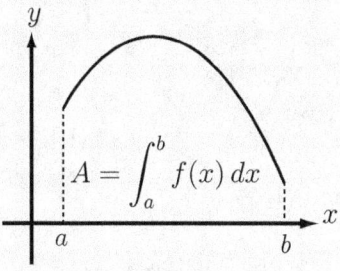

Figure 1

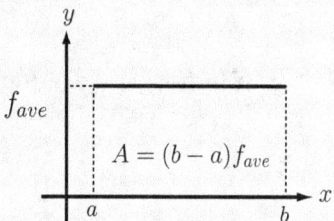

Figure 2

Example 1 Finding the average value of the function

Find the average value of the function $f(x) = \dfrac{1}{x}$ on the interval $[1,2]$.

Solution With $a = 1$ and $b = 2$,

$$f_{ave} = \frac{1}{b-a} \int_a^b f(x)\, dx = \frac{1}{2-1} \int_1^2 \frac{1}{x}\, dx$$

$$= \ln|x| \Big]_1^2 = \ln 2 - \ln 1$$

$$= \ln 2$$

Therefore, the average value of the function $f(x) = \dfrac{1}{x}$ on the interval $[1,2]$ is $\ln 2$.

MR. RHEE'S BRILLIANT MATH SERIES
AB & BC — AP CAL LESSON 25

Arc Length Formula

If f' is continuous on $[a, b]$, then the length of the curve $y = f(x)$ for $a \leq x \leq b$ is

$$L = \int_a^b \sqrt{1 + [f'(x)]^2}\, dx$$

If a curve has the equation $x = g(y)$ for $c \leq y \leq d$ and $g'(y)$ is continuous, then the length of the curve $x = g(y)$ for $c \leq y \leq d$ is

$$L = \int_c^d \sqrt{1 + [g'(y)]^2}\, dy$$

Example 2 Finding the arc length of the function

Find the arc length of the $f(x) = \dfrac{2}{3}\sqrt{x^3}$ between $(0,0)$ and $(1, \frac{2}{3})$.

Solution $f'(x) = \frac{2}{3} \cdot \frac{3}{2}x^{\frac{1}{2}} = x^{\frac{1}{2}}$. Thus, the arc length L of the function is

$$L = \int_a^b \sqrt{1 + [f'(x)]^2}\, dx = \int_0^1 \sqrt{1 + [\sqrt{x}]^2}\, dx = \int_0^1 \sqrt{1 + x}\, dx$$

Use the U-Substitution Rule to evaluate $\int_0^1 \sqrt{1+x}\, dx$. Let $u = 1 + x$. Then $du = dx$. The new lower limit and new upper limits for the integral when $x = 0$ and $x = 1$ are

When $x = 0$, $u = 1 + x = 1$,   When $x = 1$, $u = 1 + x = 2$

Thus, the new lower limit and upper limit are 1 and 2, respectively.

$$\int_0^1 \sqrt{1+x}\, dx = \int_1^2 \sqrt{u}\, du = \frac{2}{3}u^{\frac{3}{2}}\Big]_1^2$$

$$= \frac{2}{3}(2)^{\frac{3}{2}} - \frac{2}{3}(1)^{\frac{3}{2}}$$

$$= \frac{2}{3}(2\sqrt{2} - 1)$$

EXERCISES

For questions 1-4, find the average value of the following functions on the given interval.

1. $f(x) = x^3$, $[0, 2]$

2. $f(x) = \cos x$, $\left[0, \dfrac{\pi}{2}\right]$

3. $f(x) = xe^{x^2}$, $[0, 2]$

4. $f(x) = \dfrac{1}{\sqrt{1-2x}}$, $[-2, -1]$

MR. RHEE'S BRILLIANT MATH SERIES AB & BC   AP CAL LESSON 25

For questions 5-6, Set up an integral for arc length of the following functions. Do not evaluate it.

5. $f(x) = xe^x, \quad 1 \leq x \leq 3$

6. $f(x) = \ln x^2, \quad 2 \leq x \leq 5$

For questions 7-8, Set up an integral for arc length of the following functions and evaluate it using a calculator.

7. $f(x) = x^2 - 2x - 3, \quad -1 \leq x \leq 3$

8. $f(x) = e^{2x}, \quad 1 \leq x \leq 2$

MR. RHEE'S BRILLIANT MATH SERIES — AB & BC — AP CAL LESSON 25

Answers

1. 2
2. $\dfrac{2}{\pi}$
3. $\dfrac{1}{4}\left(e^4 - 1\right)$
4. $\sqrt{5} - \sqrt{3}$
5. $L = \displaystyle\int_1^3 \sqrt{1 + (e^x + xe^x)^2}\, dx$
6. $L = \displaystyle\int_2^5 \sqrt{1 + \left(\dfrac{2}{x}\right)^2}\, dx$
7. 9.294
8. 47.224

LESSON 26

Volumes Of Solids Of Revolution

Volume by the Disk Method

The region shown in Figure 1 is enclosed by $y = f(x)$ and vertical lines $x = a$ and $x = b$. If the region is rotated about the x-axis, then we get the solid shown in Figure 2.

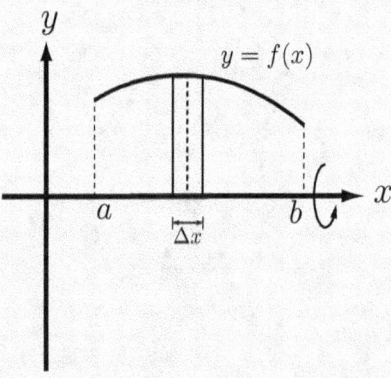

Figure 1

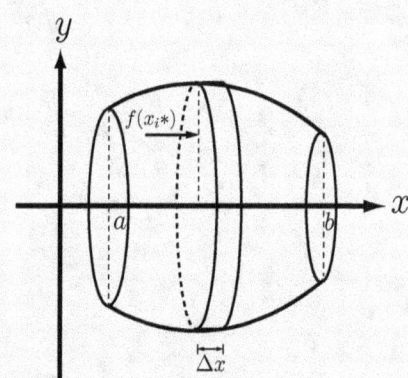

Figure 2

When we slice the solid in $[x_{i-1}, x_i]$, we get a disk with radius $f(x_i*)$ and with Δx. The volume of ith disk is

$$\pi (f(x_i*))^2 \Delta x$$

The volume V of the solid equals the sum of volumes of infinitely many disks. Thus,

$$V = \lim_{n \to \infty} \sum_{i=1}^{n} \pi [f(x_i^*)]^2 \Delta x = \pi \int_a^b [f(x)]^2 \, dx$$

Tip
1. In order to find the volume of the solid by the Disk method, draw an ith rectangle **perpendicular** to the line of rotation.

$$\text{For a vertical } i\text{th rectangle:} \implies \int_a^b dx$$

$$\text{For a horizontal } i\text{th rectangle:} \implies \int_c^d dy$$

2. If $\int dx$ is set up, then the integrand must be a function of x; that is, $\int_a^b f(x) \, dx$.

Whereas, if $\int dy$ is set up, the integrand must be a function of y; that is $\int_c^d g(y) \, dy$

MR. RHEE'S BRILLIANT MATH SERIES — AB & BC — AP CAL LESSON 26

Volume of a Solid of Revolution: Rotating about the x- or y-axis

If the region enclosed by $y = f(x)$, $x = a$, and $x = b$ as shown in Figure 3 is rotated about the x-axis, the volume of a solid is given by

$$V = \pi \int_a^b [f(x)]^2 \, dx$$

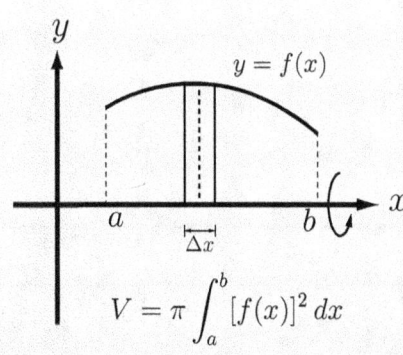

$$V = \pi \int_a^b [f(x)]^2 \, dx$$

Figure 3: Rotating about the x-axis

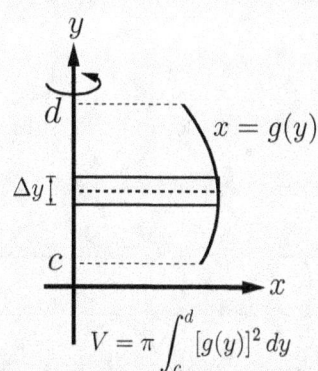

$$V = \pi \int_c^d [g(y)]^2 \, dy$$

Figure 4: Rotating about the y-axis

If the region enclosed by $x = g(y)$, $y = c$, and $y = d$ as shown in Figure 4 is rotated about the y-axis, the volume of a solid is given by

$$V = \pi \int_c^d [g(y)]^2 \, dy$$

Tip Before using the Disk method, draw an ith rectangle **perpendicular** to the line of rotation, either x- or y- axis.

Example 1 Finding the volume by the Disk method

Find the volume of the solid obtained by rotating the region bounded by $y = \sqrt{x}$ and the x-axis from 0 to 4 about the x-axis.

Solution The line of rotation is the x-axis. So, draw an ith rectangle perpendicular to the x-axis as shown below.

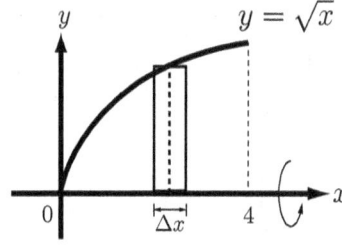

The volume of ith disk is $\pi(\sqrt{x})^2 \Delta x$ and the volume V of the solid is

$$V = \pi \int_a^b [f(x)]^2\, dx = \pi \int_0^4 (\sqrt{x})^2\, dx$$
$$= \pi \int_0^4 x\, dx = \pi \left. \frac{1}{2} x^2 \right]_0^4$$
$$= 8\pi$$

Volume by the Washer Method

The region shown in Figure 5 is enclosed by $y = f(x)$ and $y = g(x)$. Draw a vertical ith rectangle and rotate it about the x-axis as shown in Figure 6. Then we get an ith washer (a ring) with outer radius r_{out} of $f(x)$ and inner radius r_{out} of $g(x)$. The volume of the ith washer is

$$\pi\Big((r_{out})^2 - (r_{in})^2\Big) \Delta x$$

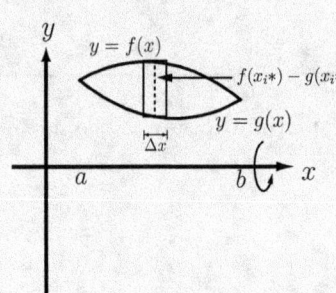
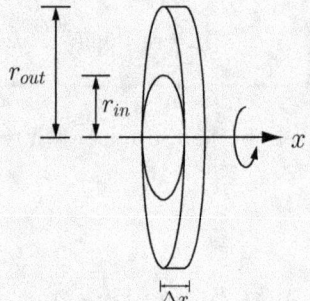
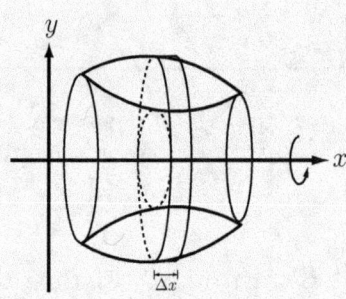

Figure 5 Figure 6 Figure 7

and the volume V of a solid shown in Figure 7 is

$$V = \pi \int_a^b \Big((r_{out})^2 - (r_{in})^2\Big) dx = \pi \int_a^b \Big((f(x))^2 - (g(x))^2\Big) dx$$

Tip — Use the Washer method when there is an open space between the enclosed region and the line of rotation. Similar to the Disk method, draw an ith rectangle **perpendicular** to the line of rotation.

MR. RHEE'S BRILLIANT MATH SERIES — AB & BC — AP CAL LESSON 26

Volume of a Solid of Revolution: Rotating about a Line $y = m$

1. The region enclosed by $y = f(x)$, $y = g(x)$, $x = a$, and $x = b$ **lies above** the line of rotation $y = m$ as shown in Figure 8. If the region is rotated about the $y = m$ as shown in Figure 9, the volume of the ith washer is $\pi\left((r_{out})^2 - (r_{in})^2\right)\Delta x$.

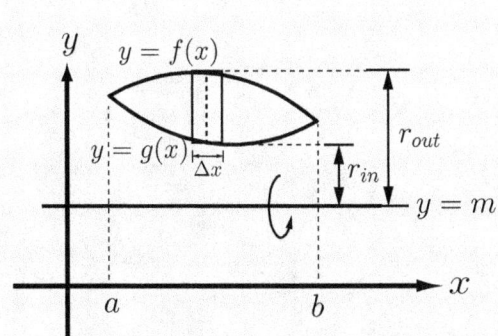

Figure 8

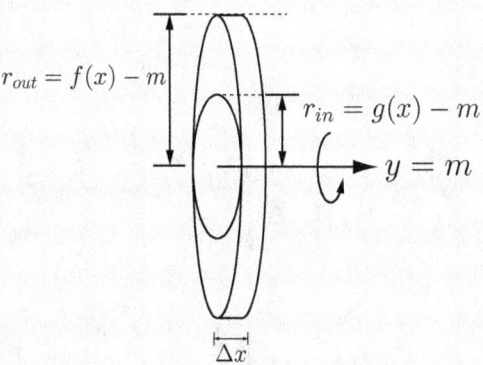

Figure 9

Thus, the volume V of the solid is $V = \pi \int_a^b \left((f(x) - m)^2 - (g(x) - m)^2\right) dx$

2. The region enclosed by $y = f(x)$, $y = g(x)$, $x = a$, and $x = b$ **lies below** the line of rotation $y = m$ as shown in Figure 10. If the region is rotated about the $y = m$ as shown in Figure 11, the volume of the ith washer is $\pi\left((r_{out})^2 - (r_{in})^2\right)\Delta x$.

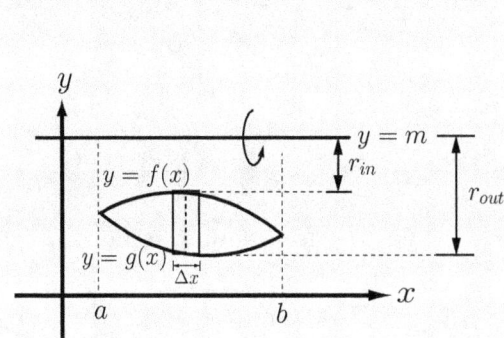

Figure 10

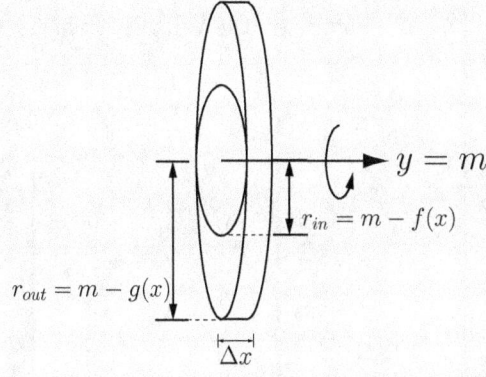

Figure 11

Thus, the volume V of the solid is $V = \pi \int_a^b \left((m - g(x))^2 - (m - f(x))^2\right) dx$

Volume of a Solid of Revolution: Rotating about a Line $x = n$

1. The region enclosed by $x = f(y)$, $x = g(y)$, $y = c$, and $y = d$ **lies to the right** of the line of rotation $x = n$ as shown in Figure 12. If the region is rotated about the $x = n$ as shown in Figure 13, the volume of the ith washer is $\pi\left((r_{out})^2 - (r_{in})^2\right)\Delta y$.

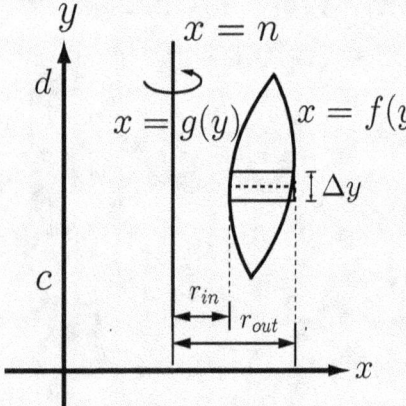

Figure 12

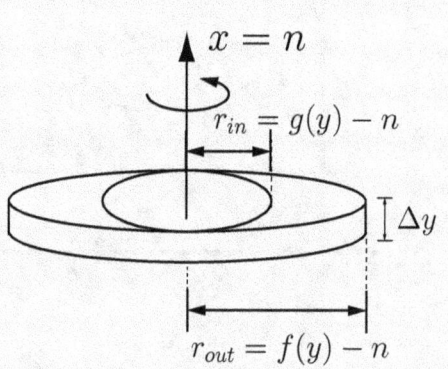

Figure 13

Thus, the volume V of the solid is $V = \pi \int_c^d \left((f(y) - n)^2 - (g(y) - n)^2\right) dy$

2. The region enclosed by $x = f(y)$, $x = g(y)$, $y = c$, and $y = d$ **lies to the left** of the line of rotation $y = m$ as shown in Figure 14. If the region is rotated about the $x = n$ as shown in Figure 15, the volume of the ith washer is $\pi\left((r_{out})^2 - (r_{in})^2\right)\Delta y$.

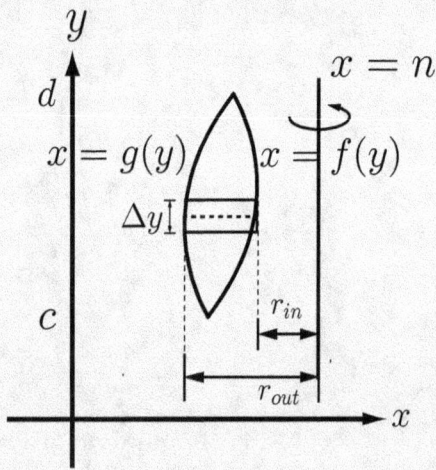

Figure 14

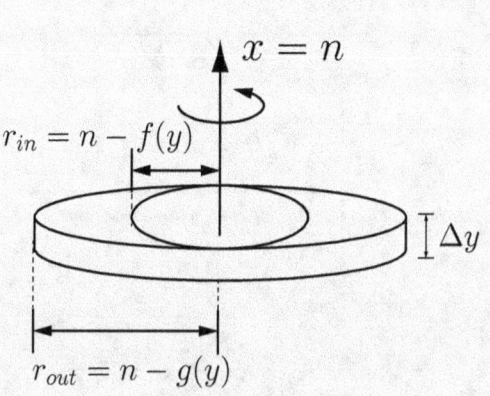

Figure 15

Thus, the volume V of the solid is $V = \pi \int_c^d \left((n - g(y))^2 - (n - f(y))^2\right) dy$

MR. RHEE'S BRILLIANT MATH SERIES   AB & BC   AP CAL LESSON 26

EXERCISES

For questions 1-6, find the volume of the solid obtained by rotating the region bounded by the given curves about the specified line.

1. $y = x^2$, $y = 2$, $x = 0$; about the y-axis

2. $y = \sqrt{x-1}$, $x = 4$, $y = 0$; about the x-axis

3. $y = -x^2 + 2x$, $y = 0$; about the x-axis

4. $y = e^x$, $y = x$, $x = 0$, $x = 1$; about the x-axis

MR. RHEE'S BRILLIANT MATH SERIES   AB & BC   AP CAL LESSON 26

5. $x = y^2 + 2$, $y = 2$, $y = -1$, $x = 1$; about the y-axis

6. $y = -x^2 + 3x$, $y = 2x$; about $y = -1$

7. The region R is enclosed by $y = \sqrt{x}$ and $y = x$. Find the volume of the solid if R is rotated about

   (a) the x-axis

   (b) $y = -2$

   (c) $y = 3$

   (d) the y-axis

   (e) $x = 5$

   (f) $x = -3$

MR. RHEE'S BRILLIANT MATH SERIES — AB & BC — AP CAL LESSON 26

Answers

1.	2π
2.	$\dfrac{9\pi}{2}$
3.	3.351
4.	8.989
5.	86.708
6.	2.199
7 (a).	$\dfrac{\pi}{6}$
7 (b).	2.618
7 (c).	2.618
7 (d).	0.419
7 (e).	4.817
7 (f).	3.560

MR. RHEE'S BRILLIANT MATH SERIES   AB & BC   AP CAL LESSON 27

LESSON 27

Volumes Of Solids Of Cross-Sections

Definition of Volume

Suppose a solid S lies between $x = a$ and $x = b$. If the cross-sectional area $A(x)$ of S is perpendicular to the x-axis, where A is a continuous function, then the volume V of S is

$$V = \int_a^b A(x)\,dx$$

Similarly, suppose a solid S lies between $y = c$ and $y = d$. If the cross-sectional area $A(y)$ of S is perpendicular to the y-axis, where A is a continuous function, then the volume V of S is

$$V = \int_c^d A(y)\,dy$$

[Tip]

1. Below summarizes the most common cross-sectional areas that are used in Calculus.

    - The area of a square with side T: $A = T^2$
    - The area of an equilateral triangle with side T: $A = \dfrac{\sqrt{3}}{4}T^2$
    - The area of an isosceles right triangle with hypotenuse T: $A = \dfrac{1}{4}T^2$
    - The area of a circle with diameter T: $A = \dfrac{\pi}{4}T^2$
    - The area of a semi-circle with diameter T: $A = \dfrac{\pi}{8}T^2$

2. If the side T in the cross sectional area formula shown above is determined by $f(x) - g(x)$, replace T by $f(x) - g(x)$. For instance, if the diameter of semi-circle is $f(x) - g(x)$, the area of semi-circle is $A = \dfrac{\pi}{8}\Big(f(x) - g(x)\Big)^2$.

MR. RHEE'S BRILLIANT MATH SERIES — AB & BC — AP CAL LESSON 27

Volumes of Solids of Cross-Sections

If the side T of cross-section is given by $f(x) - g(x)$ and the cross-sectional area $A(x)$ is written in terms of $f(x) - g(x)$, the volume V of a solid is given as follows:

- If the cross-section is a square with side $f(x) - g(x)$:

$$V = \int_a^b A(x)\,dx = \int_a^b \left(f(x) - g(x)\right)^2 dx$$

- If the cross-section is an equilateral triangle with side $f(x) - g(x)$:

$$V = \int_a^b A(x)\,dx = \frac{\sqrt{3}}{4} \int_a^b \left(f(x) - g(x)\right)^2 dx$$

- If the cross-section is an isosceles right triangle with hypotenuse $f(x) - g(x)$:

$$V = \int_a^b A(x)\,dx = \frac{1}{4} \int_a^b \left(f(x) - g(x)\right)^2 dx$$

- If the cross-section is a circle with diameter $f(x) - g(x)$:

$$V = \int_a^b A(x)\,dx = \frac{\pi}{4} \int_a^b \left(f(x) - g(x)\right)^2 dx$$

- If the cross-section is a semi-circle with diameter $f(x) - g(x)$:

$$V = \int_a^b A(x)\,dx = \frac{\pi}{8} \int_a^b \left(f(x) - g(x)\right)^2 dx$$

MR. RHEE'S BRILLIANT MATH SERIES   AB & BC   AP CAL LESSON 27

Example 1   Finding the volume of the solid of different cross-sections

The base is the region enclosed by $y = \sqrt{1-x^2}$ and the x-axis. Find the volume of solid if cross-sections perpendicular to the x-axis are

(a) squares

(b) equilateral triangles

(c) semi-circles with diameter equal to the base

Solution

(a) The cross sections are squares as shown in the figure below. Let's find a typical cross-sectional area. Since the base is the region enclosed by $y = \sqrt{1-x^2}$ and the x-axis, the side of the square is $\sqrt{1-x^2}$. Thus, the cross-sectional area is $A(x) = (\sqrt{1-x^2})^2 = 1 - x^2$.

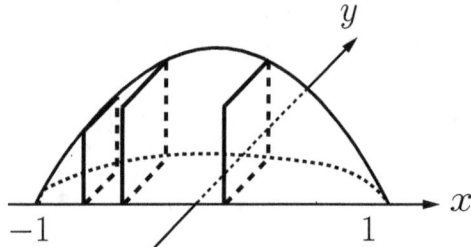

Therefore, the volume of the solid is

$$V = \int_a^b A(x)\, dx = \int_{-1}^{1} 1 - x^2\, dx = x - \frac{1}{3}x^3 \Big]_{-1}^{1} = \frac{4}{3}$$

(b) The cross sections are equilateral triangles as shown in the figure below. The cross-sectional area is $A(x) = \dfrac{\sqrt{3}}{4}(\sqrt{1-x^2})^2 = \dfrac{\sqrt{3}}{4}(1-x^2)$.

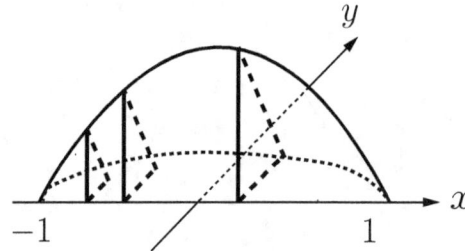

Therefore, the volume of the solid is

$$V = \int_a^b A(x)\,dx = \frac{\sqrt{3}}{4}\int_{-1}^1 1 - x^2\,dx = \frac{\sqrt{3}}{3}$$

(c) The cross sections are semi-circles as shown in the figure below. The cross-sectional area is $A(x) = \frac{\pi}{8}(\sqrt{1-x^2})^2 = \frac{\pi}{8}(1-x^2)$.

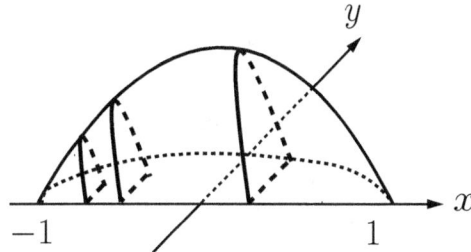

Therefore, the volume of the solid is

$$V = \int_a^b A(x)\,dx = \frac{\pi}{8}\int_{-1}^1 1 - x^2\,dx = \frac{\pi}{6}$$

EXERCISES

1. The base is a region enclosed by $y = x^2$ and $y = -x^2 + 2$.

   (a) Find the area of the base.

   (b) Find the volume of the solid if cross-sections perpendicular to the x-axis are isosceles right triangle with hypotenuse in the base.

   (c) Find the volume of the solid if cross-sections perpendicular to the x-axis are rectangles whose length is in the base and the width is twice the length.

2. The base is a region enclosed by $y = \sqrt{x}$ and $y = x^2$.

   (a) Find the area of the base.

   (b) Find the volume of the solid if cross-sections perpendicular to the x-axis are semi-circles.

   (c) Find the volume of the solid if cross-sections perpendicular to the y-axis are equilateral triangles.

3. The base is a circular region with boundary curve $x^2 + y^2 = 4$.

   (a) Find the volume of the solid if cross-sections perpendicular to the x-axis are equilateral triangles.

   (b) Find the volume of the solid if cross-sections perpendicular to the x-axis are isosceles right triangle with with hypotenuse in the base.

   (c) Find the volume of the solid if cross-sections perpendicular to the x-axis are semi-circles.

4. The base is a region enclosed by $y = x^2$ and $y = 1$.

   (a) Find the volume of the solid if cross-sections perpendicular to the y-axis are squares.

   (b) Find the volume of the solid if cross-sections perpendicular to the y-axis are isosceles right triangle with with hypotenuse in the base.

   (c) Find the volume of the solid if cross-sections perpendicular to the y-axis are semi-circles.

MR. RHEE'S BRILLIANT MATH SERIES — AB & BC — AP CAL LESSON 27

Answers

1 (a)	$\dfrac{8}{3}$	
1 (b)	$\dfrac{16}{15}$	
1 (c)	$\dfrac{128}{15}$	
2 (a)	$\dfrac{1}{3}$	
2 (b)	0.050	
2 (c)	0.056	
3 (a)	18.475	
3 (b)	10.667	
3 (c)	16.755	
4 (a)	2	
4 (b)	$\dfrac{1}{2}$	
5 (c)	$\dfrac{\pi}{4}$	

MR. RHEE'S BRILLIANT MATH SERIES　　　AB & BC　　　AP CAL LESSON 28

LESSON 28

Differential Equations

> ### Differential Equations
>
> A **differential equation** in x and y is an equation that involves x, y, and its derivatives (y', y'', etc). For instance, $y' = xy$. A function f is called a **solution** of a differential equation if the equation is satisfied when $y = f(x)$ and its derivatives are substituted into the equation.
>
> In general, solving a differential equation is to find all possible solutions called the **general solution**. Finding the general solution is very tough because there is no mathematical technique that enables us to solve all differential equations yet.
>
> However, it is much easier to find the particular solution of a differential equation that satisfies a condition of the form $y(x_0) = y_0$. The condition is called the **initial condition**. For instance, $y(0) = 1$ means that $y = 1$ when $x = 0$. **Initial-value problem** is a problem of finding a solution of the differential equation that satisfies the initial condition.

Example 1　Verifying the solution of the differential equation

Verify that $y = x^2 + 3x$ is a solution of the differential equation $x^2 y'' - 2xy' + 2y = 0$.

Solution　$y = x^2 + 3x$ gives $y' = 2x + 3$ and $y'' = 2$. Substitute y, y' and y'' into the differential equation.

$$x^2 y'' - 2xy' + 2y = 0$$
$$x^2(2) - 2x(2x + 3) + 2(x^2 + 3x) = 0$$
$$2x^2 - 4x^2 - 6x + 2x^2 + 6x = 0$$
$$4x^2 - 4x^2 - 6x + 6x = 0$$
$$0 = 0 \qquad \text{Satisfied}$$

Example 2　Verifying the solution of the differential equation

For what positive value of k does the function $y = e^{kt}$ satisfy the differential equation $y'' - 4y = 0$

Solution　$y = e^{kt}$ gives $y' = ke^{kt}$ and $y'' = k^2 e^{kt}$. Substitute y, y' and y'' into the differential equation.

$$y'' - 4y = 0$$
$$k^2 e^{kt} - 4e^{kt} = 0$$
$$e^{kt}(k^2 - 4) = 0$$
$$k^2 - 4 = 0 \qquad \text{Since } e^{kt} > 0$$
$$k = 2 \qquad \text{Since } k > 0$$

Therefore, the positive value of k for which $y = e^{kt}$ satisfy the differential equation $y'' - 4y = 0$ is 2.

Slope Field

Finding an explicit solution of a differential equation can be hard or impossible. Thus, we use a graphical approach: slope fields. The slope fields give us a visual representation of the general solution of the differential equation despite the explicit solution.

If we are given a differential equation, create a slope field by plugging in the ordered pairs (x, y) into the right side of the differential equation. Each value represents the slope of tangent line to the graph of the general solution at (x, y). For instance, $y' = x + y$ means that the slope, y', at (x, y) is the sum of the x- and y-coordinates of the point. So, the slope at $(1, 1)$ is $y' = 1 + 1 = 2$. Below shows the slopes at nine points and a slope field.

(x, y)	$(1, 1)$	$(1, 0)$	$(1, -1)$	$(0, 1)$	$(0, 0)$	$(0, -1)$	$(-1, 1)$	$(-1, 0)$	$(-1, -1)$
$y' = x + y$	2	1	0	1	0	-1	0	-1	-2

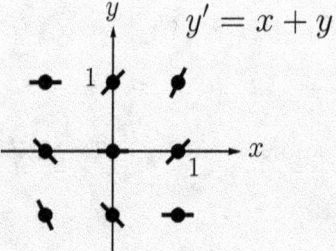

Slope Field

MR. RHEE'S BRILLIANT MATH SERIES
AB & BC — AP CAL LESSON 28

> **Separable Equations**
>
> A differential equation that can be solved explicitly or implicitly is called **separable equation**. A separable equation is a first-order differential equation in which the variables can be separated so that one variable is on side of the equation and the other variable on the other side of the equation. The following guidelines will help you solve a separable equation.
>
> **Guidelines for solving a separable equation**
>
> 1. Separate the variables: group $g(y)$ and dy on the left side, and $f(x)$ and dx on the right side.
> 2. Integrate both sides of the equation.
> 3. Write the general solution explicitly or implicitly.

Example 3 Solving the separable equation

(a) Solve the differential equation $\dfrac{dy}{dx} = \dfrac{x}{e^y + 1}$.

(b) Find the solution of this equation that satisfies the initial condition $y(1) = 0$.

Solution

(a) Let's separate the variables.

$$\frac{dy}{dx} = \frac{x}{e^y + 1}$$

$$(e^y + 1)\, dy = x\, dx \qquad \text{Integrate both sides}$$

$$\int (e^y + 1)\, dy = \int x\, dx$$

$$e^y + y = \frac{1}{2}x^2 + C$$

Thus, the general solution of $\dfrac{dy}{dx} = \dfrac{x}{e^y + 1}$ is $e^y + y = \dfrac{1}{2}x^2 + C$.

(b) In order to find the particular solution that satisfies the initial condition $y(1) = 0$, substitute $x = 1$ and $y = 0$ into the general solution obtained in (a) and solve for C.

$$e^y + y = \frac{1}{2}x^2 + C \qquad \text{Substitute } x = 1 \text{ and } y = 0$$

$$e^0 + 0 = \frac{1}{2}(1)^2 + C$$

$$1 = \frac{1}{2} + C$$

$$C = \frac{1}{2}$$

Thus, the particular solution that satisfies the initial condition $y(1) = 0$ is $e^y + y = \dfrac{1}{2}x^2 + \dfrac{1}{2}$.

MR. RHEE'S BRILLIANT MATH SERIES — AB & BC — AP CAL LESSON 28

Exponential Growth and Decay

If the rate of change in a population P at time t is proportional to the size of population, a differential equation is defined by

$$\frac{dP}{dt} = kP, \quad \text{where } k \text{ is a constant}$$

This is a separable differential equation that can be solved explicitly. The general solution of the differential equation is

$$\frac{dP}{dt} = kP \quad \Longrightarrow \quad P(t) = Ae^{kt}, \quad \text{where } A \text{ is the initial population}$$

If $k > 0$, $P(t) = Ae^{kt}$ represents exponential growth. Whereas, if $k < 0$, $P(t) = Ae^{kt}$ represents exponential decay.

Tip $\dfrac{dP}{dt} = kP$ gives $k = \dfrac{1}{P}\dfrac{dP}{dt}$, which is the growth rate divided by the population. It is called the **relative growth rate**.

Example 4 Solving the exponential growth problem

A bacteria culture starts with 100 bacteria and grows at a rate proportional to its size. After 1 hour, there are 1000 bacteria. Find the number of bacteria after 3 hours.

Solution The population of bacteria P grows at a rate proportional to its size. So, the differential equation is $\dfrac{dP}{dt} = kP$ and its general solution with initial population of 100 is $P = Ae^{kt} = 100e^{kt}$. Since the there are 1000 bacteria after 1 hour, substitute $t = 1$ and $P = 1000$ to solve for k.

$$P = 100e^{kt} \qquad \text{substitute } t = 1 \text{ and } P = 1000$$
$$1000 = 100e^{k} \qquad \text{Divide both sides by 100}$$
$$10 = e^{k} \qquad \text{Solve for } k$$
$$k = \ln 10 \approx 2.303$$

Thus, the general solution is $P(t) = 100e^{2.303t}$. In order to find the bacteria after 3 hours, substitute $t = 3$ into $P(t) = 100e^{2.303t}$.

$$P(3) = 100e^{2.303(3)} = 100,125$$

Therefore, the number of bacteria after 3 hours is $100,125$.

EXERCISES

1. Sketch a slope field for the differential equation $y' = y - x$ at the 9 points indicated.

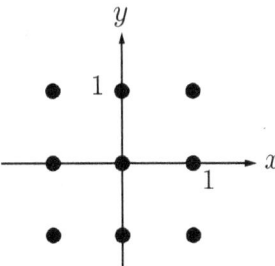

2. Solve the differential equation $\dfrac{dy}{dx} = -\dfrac{x}{y^2}$.

3. Solve the differential equation $\dfrac{dy}{dx} = \dfrac{e^{2x}}{2y}$.

4. Find the solution of the differential equation $\dfrac{dy}{dx} = x(y^2 + 1)$ that satisfies $y(1) = 0$.

MR. RHEE'S BRILLIANT MATH SERIES   AB & BC   AP CAL LESSON 28

5. Find the solution of the differential equation $\dfrac{dy}{dx} = \dfrac{\cos x}{\sin y}$ that satisfies $y\left(\dfrac{\pi}{6}\right) = \dfrac{\pi}{3}$.

6. A bacteria culture starts with 200 bacteria and grows at a rate proportional to its size. After 2 hours, there are 800 bacteria.

   (a) Find the number of bacteria after 3 hours.

   (b) Find the rate of growth after 3 hours.

7. A half-life is the amount of time it takes for half of the radioactive substance to decay. Polonium-210 has a half-life of 140 days.

   (a) If a sample has a mass of 100 mg, find the mass after 100 days.

   (b) When will the mass be reduced to 20 mg.

Answers

1.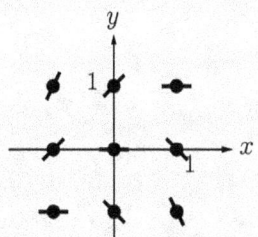

2. $\frac{1}{3}y^3 = -\frac{1}{2}x^2 + C$

3. $y^2 = \frac{1}{2}e^{2x} + C$

4. $\tan^{-1} y = \frac{1}{2}x^2 - \frac{1}{2}$

5. $-\cos y = \sin x - 1$

6 (a). 1599.29

6 (b). 1108.31

7 (a). 60.957

7 (b). 325.14

AP CACULUS

BC TOPICS

ONLY

MR. RHEE'S BRILLIANT MATH SERIES   BC ONLY   AP CAL LESSON 29

LESSON 29

Logarithmic Differentiation

Logarithmic Differentiation

For a complicated functions involving product or quotient or functions for which differentiation rules do not apply such as x^x, use **Logarithmic differentiation**.

Steps in Logarithmic Differentiation

1. Take natural logarithms of both sides of an equation and simplify using the logarithmic properties.

2. Differentiate implicitly with respect to x. Recall that whenever differentiating y, multiply the result by $\frac{dy}{dx}$.

3. Solve for y'.

Tip The logarithmic properties are as follows:

1. $\ln 1 = 0$
2. $\ln e = 1$
3. $\ln x^n = n \ln x$
4. $\ln(xy) = \ln x + \ln y$
5. $\ln\left(\dfrac{x}{y}\right) = \ln x - \ln y$

Example 1 Applying logarithmic differentiation

Differentiate $y = x^x$

Solution Take natural logarithms of both sides of the equation and simplify using the logarithmic properties.

$$y = x^x$$
$$\ln y = \ln x^x$$
$$\ln y = x \ln x$$

Differentiate y implicitly with respect to x and solve for $\frac{dy}{dx}$.

$$\frac{1}{y}\frac{dy}{dx} = \ln x + 1$$

$$\frac{dy}{dx} = y(\ln x + 1)$$

$$\frac{dy}{dx} = x^x(\ln x + 1)$$

Therefore, the derivative of $y = x^x$ is $\frac{dy}{dx} = x^x(\ln x + 1)$.

[Tip] Recall that

- Power rule: $(x^n)' = nx^{n-1}$, where the base is variable and exponent is constant.
- Rule for differentiating exponential function: $(a^x)' = a^x \ln a$, where the base is constant and exponent is variable.

Thus, you cannot apply the power rule to differentiate the function $y = x^x$, where the base is variable and exponent is variable. Simply put, $y \neq x(x)^{x-1}$.

Example 2 Applying logarithmic differentiation

Differentiate $y = \dfrac{x\sqrt{x^2+1}}{(1-2x)^2}$.

Solution Take natural logarithms of both sides of the equation and simplify using the logarithmic properties.

$$\ln y = \ln \frac{x\sqrt{x^2+1}}{(1-2x)^2}$$

$$\ln y = \ln x + \frac{1}{2}\ln(x^2+1) - 2\ln(1-2x)$$

Differentiating y implicitly with respect to x gives

$$\frac{1}{y}\frac{dy}{dx} = \frac{1}{x} + \frac{1}{2} \cdot \frac{1}{x^2+1} \cdot (2x) - 2 \cdot \frac{1}{1-2x} \cdot (-2)$$

$$\frac{1}{y}\frac{dy}{dx} = \frac{1}{x} + \frac{x}{x^2+1} + \frac{4}{1-2x}$$

Solving for $\frac{dy}{dx}$, we get

$$\frac{dy}{dx} = y\left(\frac{1}{x} + \frac{x}{x^2+1} + \frac{4}{1-2x}\right)$$

$$\frac{dy}{dx} = \frac{x\sqrt{x^2+1}}{(1-2x)^2}\left(\frac{1}{x} + \frac{x}{x^2+1} + \frac{4}{1-2x}\right)$$

EXERCISES

For questions 1-6, use the logarithmic differentiation to find the derivative of the following functions.

1. $y = (3x-1)^5(x^2+2x-1)^6$

2. $y = \dfrac{e^{x^2}(x^4+1)}{\sqrt{x-1}}$

3. $y = \sqrt[3]{\dfrac{x^3-1}{x^3+1}}$

4. $y = x^{\sqrt{x}}$

5. $y = x^{\sin x}$

6. $y = (\ln x)^x$

MR. RHEE'S BRILLIANT MATH SERIES   BC ONLY   AP CAL LESSON 29

Answers

1. $\dfrac{dy}{dx} = (3x-1)^5(x^2+2x-1)^6\left(\dfrac{15}{3x-1} + \dfrac{12(x+1)}{x^2+2x-1}\right)$

2. $\dfrac{dy}{dx} = \dfrac{e^{x^2}(x^4+1)}{\sqrt{x-1}}\left(2x + \dfrac{4x^3}{x^4+1} - \dfrac{1}{2(x-1)}\right)$

3. $\dfrac{dy}{dx} = \sqrt[3]{\dfrac{x^3-1}{x^3+1}}\left(\dfrac{2x^2}{x^3-1} \cdot \dfrac{1}{x^3+1}\right)$

4. $\dfrac{dy}{dx} = x^{\sqrt{x}}\left(\dfrac{\ln x}{2\sqrt{x}} + \dfrac{\sqrt{x}}{x}\right)$

5. $\dfrac{dy}{dx} = x^{\sin x}\left(\cos x \ln x + \dfrac{\sin x}{x}\right)$

6. $\dfrac{dy}{dx} = (\ln x)^x\left(\ln(\ln x) + \dfrac{1}{\ln x}\right)$

MR. RHEE'S BRILLIANT MATH SERIES   BC ONLY   AP CAL LESSON 30

LESSON 30

Indeterminate Products and Indeterminate Powers

L'Hospital's Rule

Suppose you have indeterminate forms of

$$\lim_{x \to a} \frac{f(x)}{g(x)} = \frac{0}{0} \quad \text{or} \quad \lim_{x \to a} \frac{f(x)}{g(x)} = \frac{\pm\infty}{\pm\infty}$$

Then,

$$\lim_{x \to a} \frac{f(x)}{g(x)} = \lim_{x \to a} \frac{f'(x)}{g'(x)}$$

Tip
1. L'Hospital's Rule says that the limit of a quotient of functions is equal to the limit of the quotient of their derivatives.

2. L'Hospital's Rule is also valid for one-sided limits: that is, $x \to a$ can be replaced by any of the followings: $x \to a^+$, $x \to a^-$.

Indeterminate Products

Suppose $\lim_{x \to a} f(x) = 0$ and $\lim_{x \to a} g(x) = \infty$. Then $\lim_{x \to a} f(x)g(x) = 0 \cdot \infty$. This kind of limit is called an **Indeterminate form of type $0 \cdot \infty$**. The L'Hospital's Rule does not work on products, it only works on quotients.

Steps in finding the limit of an indeterminate form of type $0 \cdot \infty$

1. Rewrite the product fg as a quotient shown below.

$$fg = \frac{f}{\frac{1}{g}} \quad \text{or} \quad fg = \frac{g}{\frac{1}{f}}$$

Then

$$\lim_{x \to a} fg = \lim_{x \to a} \frac{f}{\frac{1}{g}} = \frac{0}{0} \quad \text{or} \quad \lim_{x \to a} fg = \lim_{x \to a} \frac{g}{\frac{1}{f}} = \frac{\infty}{\infty}$$

the indeterminate form of type $0 \cdot \infty$ changes to an indeterminate form of $\frac{0}{0}$ or $\frac{\infty}{\infty}$.

2. Use L'Hospital's Rule.

Tip
When you rewrite the product fg as a quotient either $fg = \frac{f}{1/g}$ or $fg = \frac{g}{1/f}$, select one that leads you to the simpler limit after you apply L'Hospital's Rule. If you get a more complicated expression than the one you started with, select the other quotient.

MR. RHEE'S BRILLIANT MATH SERIES   **BC ONLY**   **AP CAL LESSON 30**

Example 1   Finding the limit of an indeterminate product

Evaluate $\lim\limits_{x \to 0^+} x \ln x$.

Solution   Plugging-in $x = 0^+$ into $x \ln x$, we get an indeterminate form of $0 \cdot \infty$. Rewrite the product $x \ln x$ as $\dfrac{\ln x}{1/x}$ so that the indeterminate form of $0 \cdot \infty$ becomes an indeterminate form of $\dfrac{\infty}{\infty}$. Use L'Hospital's Rule to evaluate the limit.

$$\begin{aligned}
\lim_{x \to 0^+} x \ln x &= \lim_{x \to 0^+} \frac{\ln x}{\frac{1}{x}} &&\text{Use L'Hospital's Rule} \\
&= \lim_{x \to 0^+} \frac{\frac{1}{x}}{-\frac{1}{x^2}} \\
&= \lim_{x \to 0^+} (-x) \\
&= 0
\end{aligned}$$

Tip   In case you rewrite the product $x \ln x$ as a quotient, $\dfrac{x}{1/\ln x}$, and apply L'Hospital's Rule,

$$\begin{aligned}
\lim_{x \to 0^+} x \ln x &= \lim_{x \to 0^+} \frac{x}{\frac{1}{\ln x}} &&\text{Use L'Hospital's Rule} \\
&= \lim_{x \to 0^+} \frac{1}{-(\ln x)^{-2} \cdot \frac{1}{x}} \\
&= \lim_{x \to 0^+} -x(\ln x)^2
\end{aligned}$$

you get a more complicated expression than the one started with. So, select other quotient.

Indeterminate Differences

Suppose $\lim\limits_{x \to a} f(x) = \infty$ and $\lim\limits_{x \to a} g(x) = \infty$. Then $\lim\limits_{x \to a} [f(x) - g(x)] = \infty - \infty$. This kind of limit is called an **Indeterminate form of type** $\infty - \infty$. L'Hospital's Rule does not work on differences, it only works on quotients. Thus, rewrite the difference as a quotient and use L'Hospital's Rule.

MR. RHEE'S BRILLIANT MATH SERIES — BC ONLY — AP CAL LESSON 30

Example 2 Finding the limit of an indeterminate difference

Evaluate $\lim\limits_{x \to \infty}(xe^{\frac{1}{x}} - x)$.

Solution

Plugging-in $x = \infty$ into $xe^{\frac{1}{x}} - x$, we get an indeterminate form of $\infty - \infty$. Rewrite the difference as a quotient such that $xe^{\frac{1}{x}} - x = x(e^{\frac{1}{x}} - 1) = \dfrac{e^{\frac{1}{x}} - 1}{\frac{1}{x}}$.

$$\lim_{x \to \infty}(xe^{\frac{1}{x}} - x) = \lim_{x \to \infty}\frac{e^{\frac{1}{x}} - 1}{\frac{1}{x}} = \frac{0}{0}$$

Thus, the indeterminate form of $\infty - \infty$ becomes an indeterminate form of $\frac{0}{0}$. Use L'Hospital's Rule to evaluate the limit.

$$\lim_{x \to \infty} \frac{e^{\frac{1}{x}} - 1}{\frac{1}{x}} = \lim_{x \to \infty} \frac{e^{\frac{1}{x}}\left(-\frac{1}{x^2}\right)}{-\frac{1}{x^2}}$$
$$= \lim_{x \to \infty} e^{\frac{1}{x}}$$
$$= 1$$

Indeterminate Powers

If $\lim\limits_{x \to a}[f(x)]^{g(x)}$ has the following indeterminate power forms,

$$0^0 \qquad \infty^0 \qquad 1^\infty$$

let $y = \lim\limits_{x \to a}[f(x)]^{g(x)}$ and takes the natural logarithm on both sides of the equation so that the indeterminate power form becomes an indeterminate product form.

$$y = \lim_{x \to a}[f(x)]^{g(x)} \quad \Longrightarrow \quad \ln y = \lim_{x \to a} g(x) \ln f(x) = 0 \cdot \infty$$

Then, rewrite the product as a quotient and use L'Hospital's Rule.

Tip Notice that 0^∞, ∞^∞, 1^0 are **NOT** indeterminate power forms.

Example 3 Finding the limit of an indeterminate power

Evaluate $\lim\limits_{x \to 0^+} x^x$.

Solution Plugging-in $x = 0^+$ into x^x, we get an indeterminate power form of 0^0. Let $y = \lim\limits_{x \to 0^+} x^x$ and take the natural logarithm on both sides of the equation.

$$y = \lim_{x \to 0^+} x^x$$
$$\ln y = \lim_{x \to 0^+} x \ln x$$

Since $\lim\limits_{x \to 0^+} x \ln x = 0$ shown in the example 1,

$$\ln y = 0$$
$$y = e^0 = 1$$

Therefore, $y = \lim\limits_{x \to 0^+} x^x = 1$.

MR. RHEE'S BRILLIANT MATH SERIES   BC ONLY   AP CAL LESSON 30

EXERCISES

1. Evaluate $\lim\limits_{x \to \infty} e^{-x} \ln x$.

2. Evaluate $\lim\limits_{x \to \infty} x^2 e^{-x}$.

3. Evaluate $\lim\limits_{x \to \infty} x^{\frac{1}{x}}$.

4. Evaluate $\lim\limits_{x \to 0} (1+x)^{\frac{1}{x}}$.

5. Evaluate $\lim_{x \to \infty} (x - \sqrt{x^2 - 1})$.

6. Evaluate $\lim_{x \to 0^+} x^{\sin x}$.

MR. RHEE'S BRILLIANT MATH SERIES
BC ONLY   AP CAL LESSON 30

Answers

1. 0   2. 0   3. 1
4. e 5. 0   6. 1

MR. RHEE'S BRILLIANT MATH SERIES   BC ONLY   AP CAL LESSON 31

LESSON 31

Derivative And Arc Length Of Parametric Equations

Derivative of Parametric Equations

If a curve C is defined by parametric equations $x = f(t)$ and $y = g(t)$, then the slope of the tangent line to the curve C at (x, y) is given by

$$\frac{dy}{dx} = \frac{\frac{dy}{dt}}{\frac{dx}{dt}}, \quad \text{where } \frac{dx}{dt} \neq 0$$

and the second derivative $\dfrac{d^2y}{dx^2}$ is given by

$$\frac{d^2y}{dx^2} = \frac{d}{dx}\left(\frac{dy}{dx}\right) = \frac{\frac{d}{dt}\left(\frac{dy}{dx}\right)}{\frac{dx}{dt}}$$

Tip

1. The curve C defined by parametric equations has a horizontal tangent when $\dfrac{dy}{dt} = 0$ and it has a vertical tangent when $\dfrac{dx}{dt} = 0$

2. $\dfrac{dy}{dx} = \dfrac{dy}{dt} / \dfrac{dx}{dt}$ enables us to find the slope the tangent line to the curve C without having to eliminate the parameter t.

Arc Length of Parametric Equations

If a curve C is defined by parametric equations $x = f(t)$ and $y = g(t)$, where $\alpha \leq t \leq \beta$, then the arc length L of the curve C is given by

$$L = \int_{\alpha}^{\beta} \sqrt{\left(\frac{dx}{dt}\right)^2 + \left(\frac{dy}{dt}\right)^2}\, dt$$

Tip   Recall that the arc length of the curve $y = f(x)$ for $a \leq x \leq b$ is

$$L = \int_{a}^{b} \sqrt{1 + \left(\frac{dy}{dx}\right)^2}\, dx$$

MR. RHEE'S BRILLIANT MATH SERIES BC ONLY AP CAL LESSON 31

Example 1 Finding the slope and arc length of parametric equations

Consider a curve C defined by the parametric equations $x(t) = 2\sin t$, $y(t) = 2\cos t$, where $0 \le t \le \pi$.

(a) Find an equation of the tangent line to the curve at $t = \dfrac{\pi}{4}$.

(b) Find the arc length of the curve C.

Solution

(a) $\dfrac{dy}{dt} = -2\sin t$ and $\dfrac{dx}{dt} = 2\cos t$.

$$\frac{dy}{dx} = \frac{\frac{dy}{dt}}{\frac{dx}{dt}} = \frac{-2\sin t}{2\cos t} = -\tan t$$

Thus, the slope of the tangent to the curve at $t = \dfrac{\pi}{4}$ is

$$\left.\frac{dy}{dx}\right|_{t=\frac{\pi}{4}} = -\tan\left(\frac{\pi}{4}\right) = -1$$

When $t = \dfrac{\pi}{4}$,

$$x\left(\frac{\pi}{4}\right) = 2\sin\left(\frac{\pi}{4}\right) = \sqrt{2}, \qquad y\left(\frac{\pi}{4}\right) = 2\cos\left(\frac{\pi}{4}\right) = \sqrt{2}$$

Therefore, the equation of the tangent line to the curve C at $(\sqrt{2}, \sqrt{2})$ is $y - \sqrt{2} = -(x - \sqrt{2})$.

(b) Since $\dfrac{dy}{dt} = -2\sin t$ and $\dfrac{dx}{dt} = 2\cos t$,

$$L = \int_\alpha^\beta \sqrt{\left(\frac{dx}{dt}\right)^2 + \left(\frac{dy}{dt}\right)^2}\, dt$$

$$= \int_0^\pi \sqrt{(2\cos t)^2 + (-2\sin t)^2}\, dt$$

$$= \int_0^\pi \sqrt{4(\cos^2 t + \sin^2 t)}\, dt \qquad \text{Since } \cos^2 t + \sin^2 t = 1$$

$$= \int_0^\pi 2\, dt$$

$$= 2t \Big]_0^\pi$$

$$= 2\pi$$

MR. RHEE'S BRILLIANT MATH SERIES
BC ONLY   AP CAL LESSON 31

Vector Functions

If a path that a particle moves in the xy-plane is defined by the parametric equations $x = f(t)$ and $y = g(t)$, the position vector, the velocity vector, and the acceleration vector at any time t are given as follows:

$$\text{Position vector:} \quad <x(t), g(t)>$$
$$\text{Velocity vector:} \quad <x'(t), g'(t)>$$
$$\text{Accelation vector:} \quad <x''(t), g''(t)>$$

The speed of the particle or the magnitude of the velocity vector is given by

$$\text{Speed of the particle} = \sqrt{((x'(t))^2 + (y'(t))^2} = \sqrt{\left(\frac{dx}{dt}\right)^2 + \left(\frac{dy}{dt}\right)^2}$$

The distance traveled by the particle from $t = t_1$ to $t = t_2$ or the arc length of the path $t = t_1$ to $t = t_2$ is given by

$$L = \int_{t_1}^{t_2} \sqrt{\left(\frac{dx}{dt}\right)^2 + \left(\frac{dy}{dt}\right)^2} \, dt$$

Tip   From the velocity vector $<x'(t), g'(t)>$, $x'(t) = \dfrac{dx}{dt}$ is the velocity of the particle in the horizontal direction and $y'(t) = \dfrac{dy}{dt}$ is the the velocity of the particle in the vertical direction.

Example 2   Finding velocity and acceleration vectors

A particle moves in the xy-plane so that any time t, the position of the particle is given by $x(t) = t^3 + 2t + 1$, $y(t) = t^2 + 3t$.

(a) Find the velocity vector when $t = 1$.

(b) Find the speed of the particle when $t = 1$.

(c) Find the acceleration vector when $t = 2$.

Solution

(a) Since $x'(t) = 3t^2 + 2$ and $y'(t) = 2t + 3$, the velocity vector is $v(t) = <3t^2 + 2, 2t + 3>$. Thus, the velocity vector when $t = 1$ is $v(1) = <3(1)^2 + 2, 2(1) + 3> = <5, 5>$.

(b) The velocity vector when $t = 1$ is $<5, 5>$. Thus, the speed of the particle is $\sqrt{5^2 + 5^2} = 5\sqrt{2}$.

(c) Since $x''(t) = 6t$ and $y''(t) = 2$, the acceleration vector is $a(t) = <6t, 2>$, Thus, the acceleration vector when $t = 2$ is $a(2) = <12, 2>$.

EXERCISES

1. Consider a curve defined by the parametric equations $x(t) = t^2 - t$, $y(t) = \frac{1}{3}t^3 + 2t^2 + 3t$, $-5 \le t \le 5$.

   (a) Find $\dfrac{dy}{dx}$.

   (b) Find an equation of the tangent line at $t = 1$.

   (c) Find $\dfrac{d^2y}{dx^2}$.

   (d) Find the points on the curve where the tangent line is horizontal.

   (e) Find the points on the curve where the tangent line is vertical.

   (f) Set up, but do not evaluate, an integral that represents the length of the curve.

2. A particle moves in the xy-plane. Its position vector is $< \ln(2t-1), t^3 >$ for $1 \le t \le 5$.

   (a) Find its velocity vector at $t = 3$.

   (b) Find the speed of the particle at $t = 3$.

   (c) Find its acceleration vector at $t = 2$.

   (d) Set up, but do not evaluate, an integral that represents the distance traveled by the particle.

MR. RHEE'S BRILLIANT MATH SERIES   BC ONLY   AP CAL LESSON 31

3. A particle moves in the xy-plane so that its position vector is $< t^3-12t, 2t^3-9t^2+12t >$. For what values of t is the particle at rest?

4. A particle moves along the curve $xy = 12$. If $x = 3$ and $\dfrac{dy}{dt} = 2$, find the value of $\dfrac{dx}{dt}$.

Answers

1(a). $\dfrac{t^2 + 4t + 3}{2t - 1}$

1(b). $y - \dfrac{16}{3} = 8(x - 0)$

1(c). $\dfrac{2(t^2 - t - 5)}{(2t - 1)^3}$

1(d). $\left(2, -\dfrac{4}{3}\right)$, $(12, 0)$

1(e). $\left(-\dfrac{1}{4}, \dfrac{49}{24}\right)$

1(f). $\displaystyle\int_{-5}^{5} \sqrt{(2t-1)^2 + (t^2 + 4t + 3)^2}\, dt$

2(a). $<\dfrac{2}{5}, 27>$

2(b). 27.003

2(c). $<-\dfrac{4}{9}, 12>$

2(d). $\displaystyle\int_{1}^{5} \sqrt{\left(\dfrac{2}{2t-1}\right)^2 + (3t^2)^2}\, dt$

3. $t = 2$

4. -1.5

MR. RHEE'S BRILLIANT MATH SERIES   BC ONLY   AP CAL LESSON 32

LESSON 32

Volumes By Cylindrical Shells

Volume by the Shell Method

Another way of finding the volumes of solids of revolution is the **Shell** method. It enables us to find the volumes that are difficult to evaluate using the Disk or Washer method.

Figure 1 shows an ith cylindrical shell with inner radius r_1, outer radius r_2, and height h.

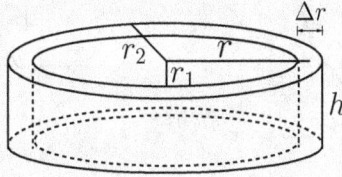

Figure 1

Then the volume V of the ith shell can be defined as $V = 2\pi rh\Delta r$, where $r = \frac{1}{2}(r_1 + r_2)$ (the average radius of the shell) and $\Delta r = r_2 - r_1$ (the thickness of the shell).

In general, the volume of a solid by the Shell method can be defined as

$$V = 2\pi \int_a^b (\text{shell radius}) \cdot (\text{shell height}) \, dr, \quad \text{where } dr = dx \text{ or } dr = dy$$

[Tip]

1. As you select the Disk or Washer method or the Shell method to find the volume of a solid, draw an ith rectangle accordingly.

   - For the Disk or Washer method: draw an ith rectangle **perpendicular** to the line of rotation.
   - For the Shell method: draw an ith rectangle **parallel** to the line of rotation.

   For a vertical ith rectangle: $\implies$ $\int_a^b dx$

   For a horizontal ith rectangle: $\implies$ $\int_c^d dy$

2. If $\int dx$ is set up, then the integrand must be a function of x; that is, $\int_a^b f(x)\, dx$.

   Whereas, if $\int dy$ is set up, the integrand must be a function of y; that is $\int_c^d g(y)\, dy$

MR. RHEE'S BRILLIANT MATH SERIES — BC ONLY — AP CAL LESSON 32

Volume of a Solid of Revolution: Rotating about the x- or y-axis

If the region enclosed by $y = f(x)$, $y = g(x)$, $x = a$, and $x = b$ as shown in Figure 2 is rotated about the y-axis, the volume of a solid is given by

$$V = 2\pi \int_a^b (\text{shell radius}) \cdot (\text{shell height}) \, dr = 2\pi \int_a^b x\Big((f(x) - g(x))\Big) \, dx$$

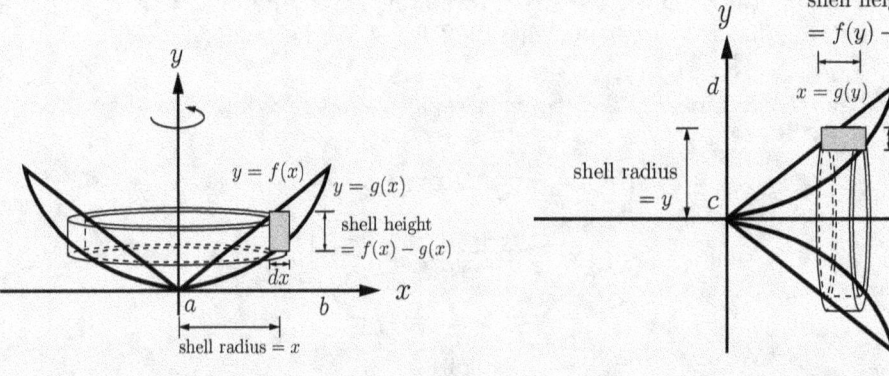

Figure 2   Figure 3

If the region enclosed by $x = f(y)$, $x = g(y)$, $y = c$, and $y = d$ as shown in Figure 3 is rotated about the x-axis, the volume of a solid is given by

$$V = 2\pi \int_c^d (\text{shell radius}) \cdot (\text{shell height}) \, dr = 2\pi \int_c^d y\Big((f(y) - g(y))\Big) \, dy$$

Tip   For the Shell method: draw an ith rectangle **parallel** to the line of rotation.

Example 1   Finding the volume by the Shell method

Use the Shell method to find the volume of the solid obtained by rotating the region enclosed by $y = \sqrt{x}$, the x-axis, and $x = 1$ about the y-axis.

Solution   Draw an ith rectangle parallel to the y-axis as shown below. Since shell radius is x and shell height is $\sqrt{x}$, the volume of solid is

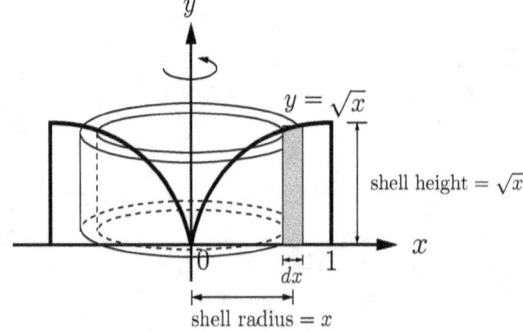

210

$$V = 2\pi \int_a^b (\text{shell radius}) \cdot (\text{shell height})\, dr$$

$$= 2\pi \int_0^1 x(\sqrt{x})\, dx = 2\pi \frac{2}{5} x^{\frac{5}{2}} \Big|_0^1 = \frac{4\pi}{5}$$

How to determine the shell radius: Rotating about a Line $x = P$

The most hardest part of setting up an integral for the volume of the solid by the Shell method is to find the shell radius, which is the distance between an ith rectangle and a line of rotation. Below shows how to determine the shell radius when the region is rotated about the line $x = P$.

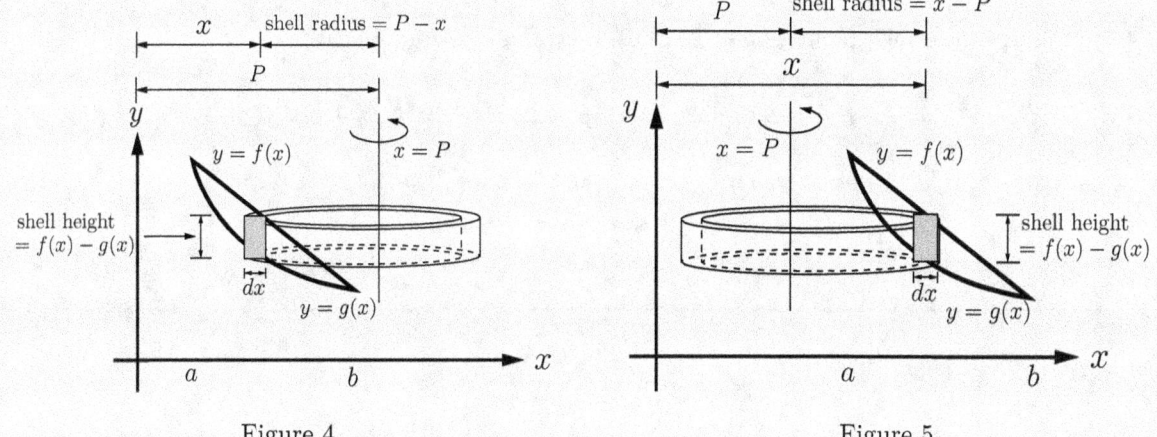

Figure 4　　　　　　　　　　　　　　Figure 5

In Figure 4, the shell radius is $P - x$. Thus, the volume of the solid is

$$V = 2\pi \int_a^b (\text{shell radius}) \cdot (\text{shell height})\, dr = 2\pi \int_a^b (P - x)\Big(f(x) - g(x)\Big)\, dx$$

In Figure 5, the shell radius is $x - P$. Thus, the the volume of the solid is

$$V = 2\pi \int_a^b (\text{shell radius}) \cdot (\text{shell height})\, dr = 2\pi \int_a^b (x - P)\Big(f(x) - g(x)\Big)\, dx$$

MR. RHEE'S BRILLIANT MATH SERIES — BC ONLY — AP CAL LESSON 32

How to determine the shell radius: Rotating about a Line $y = Q$

Below shows how to determine the shell radius when the region is rotated about the line $y = Q$.

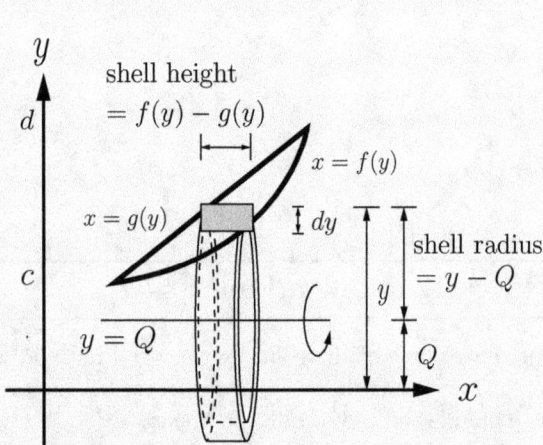
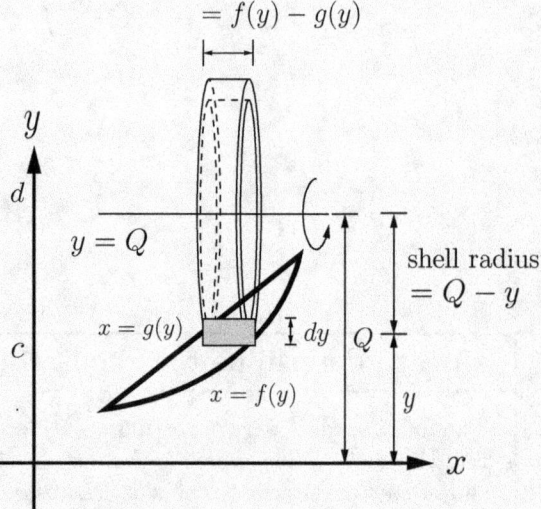

Figure 6   Figure 7

In Figure 6, the shell radius is $y - Q$. Thus, the volume of the solid is

$$V = 2\pi \int_c^d (\text{shell radius}) \cdot (\text{shell height})\, dr = 2\pi \int_a^b (y - Q)\Big(f(y) - g(y)\Big)\, dy$$

In Figure 7, the shell radius is $Q - y$. Thus, the the volume of the solid is

$$V = 2\pi \int_c^d (\text{shell radius}) \cdot (\text{shell height})\, dr = 2\pi \int_a^b (Q - y)\Big(f(y) - g(y)\Big)\, dy$$

Example 2   Find the volume of the solid by the Shell and Washer methods

A region is enclosed by $y = \sin x$ and the x-axis for $0 \le x \le \pi$. A solid is obtained by rotating the region about the y-axis.

(a) Set up, but do not evaluate, an integral for the volume of the solid by the Shell method.

(b) Set up, but do not evaluate, an integral for the volume of the solid by the Washer method.

Solution

(a) Draw an ith rectangle parallel to the y-axis as shown the figure below. Since the ith rectangle is vertical, use dx.

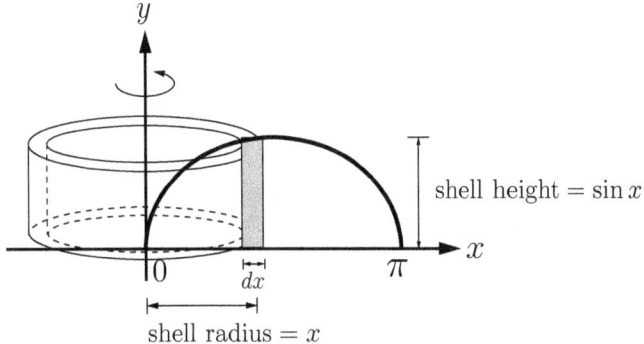

The shell radius is x and shell height is $\sin x$. Thus, the volume of the solid by the Shell method is

$$V = 2\pi \int_a^b (\text{shell radius}) \cdot (\text{shell height}) \, dr = 2\pi \int_0^\pi x \sin x \, dx$$

(b) Draw an ith rectangle parallel to the y-axis as shown the figure below. Since the ith rectangle is horizontal, use dy.

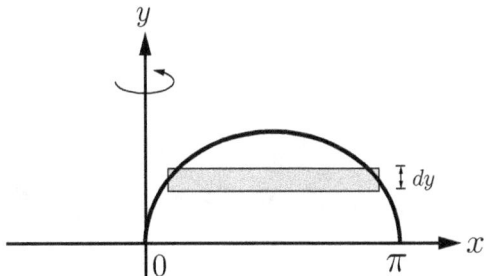

In order to use the Washer method, we need to rewrite $y = \sin x$ as $x = f(y)$ so that we can find the outer radius and inner radius of an ith washer. However, this cannot be easily done. Therefore, we cannot find the volume of the solid by the Washer method.

MR. RHEE'S BRILLIANT MATH SERIES

BC ONLY AP CAL LESSON 32

EXERCISES

For questions 1-6, use the Shell method to find the volume of the solid obtained by rotating the region enclosed by the given curves about the specified axis. Set up, but do not evaluate, an integral for the volume of the solid.

1. $y = \dfrac{1}{x}$, $x = 1$, $x = 2$, $y = 0$; about the y-axis

2. $y = x^3$, $y = 0$, $x = 2$; about $x = 4$

3. $y = x^2$, $y = 2x$; about the x-axis

4. $y = e^{-x^2}$, $x = 1$, $x = 3$, $y = 0$; about the y-axis

5. $y = \sqrt{x-1}$, $x = 5$, $y = 0$; about $y = 4$

6. $x = y^2$, $y = x^2$; about $y = -3$

MR. RHEE'S BRILLIANT MATH SERIES
BC ONLY   AP CAL LESSON 32

Answers

1. $V = 2\pi \int_{1}^{2} x \cdot \dfrac{1}{x}\, dx$

2. $V = 2\pi \int_{0}^{2} (4-x)x^3 \, dx$

3. $V = 2\pi \int_{0}^{4} y\left(\sqrt{y} - \dfrac{y}{2}\right) dy$

4. $V = 2\pi \int_{1}^{3} x e^{-x^2} \, dx$

5. $V = 2\pi \int_{0}^{2} (4-y)(4-y^2) \, dy$

6. $V = 2\pi \int_{0}^{1} (y+3)(\sqrt{y} - y^2) \, dy$

MR. RHEE'S BRILLIANT MATH SERIES BC ONLY   AP CAL LESSON 33

LESSON 33

Integration By Parts

Integration by Parts

We have evaluated so many indefinite integrals using the basic indefinite integrals shown below and the most important method of integration, the U-Substitution Rule.

1. $\int x^n = \dfrac{1}{n+1}x^{n+1} + C \quad (n \neq -1)$
2. $\int \dfrac{1}{x} = \ln|x| + C$
3. $\int a^x \, dx = \dfrac{a^x}{\ln a} + C$
4. $\int e^x \, dx = e^x + C$
5. $\int \sin x \, dx = -\cos x + C$
6. $\int \cos x \, dx = \sin x + C$
7. $\int \sec^2 x \, dx = \tan x + C$
8. $\int \csc^2 x \, dx = -\cot x + C$
9. $\int \sec x \tan x \, dx = \sec x + C$
10. $\int \csc x \cot x \, dx = -\csc x + C$
11. $\int \dfrac{1}{x^2+1} \, dx = \tan^{-1} x + C$
12. $\int \dfrac{1}{\sqrt{1-x^2}} \, dx = \sin^{-1} x + C$
13. $\int \tan x \, dx = \ln|\sec x| + C$
14. $\int \cot x \, dx = \ln|\sin x| + C$

U-Substitution Rule

$$\int f\bigl(g(x)\bigr) g'(x) \, dx = \int f(u) \, du$$

where $\int f(u) \, du$ becomes one the basic indefinite integrals shown above.

However, using the basic indefinite integrals and the U-Substitution Rule, we are still not able to evaluate an integral such as

$$\int x e^x \, dx$$

In order to find this integral, we use a method called the **Integration by parts**, which is a handy tool to find the antiderivative of functions resulted from the Product rule.

MR. RHEE'S BRILLIANT MATH SERIES
BC ONLY   AP CAL LESSON 33

Integration by parts

$$\int u\,dv = uv - \int v\,du$$

Applying integration by parts transforms a difficult integral (on the left side) into the difference of the product of two functions (in the middle) and a easy integral (on the right side)

Guidelines for choosing u and dv

$$u \xleftarrow{\quad\quad} \underset{\text{easier to differentiate}}{\ln x \quad (\sin^{-1} x, \tan^{-1} x) \quad (x^n, 1)} \quad \underset{\text{easier to integrate}}{(\sin x, \cos x) \quad e^x} \xrightarrow{\quad\quad} dv$$

A list above shows five different types of functions: a logarithmic function, an inverse trigonometric function, power function including 1, trigonometric function, and exponential function. If the integrand is a product of two functions from the list, choose one function as u if it is closer to the left side and then choose another function as dv. For instance, in $\int xe^x\,dx$, the integrand is xe^x, which is the product of a power function and an exponential function. Choose x as u and $e^x\,dx$ as dv since the power function is closer to the left side than the exponential function.

Tip   In general, use the integration by parts if the integrand is a product of two functions from the list shown above.

Example 1   Evaluating the integral using integration by parts

Evaluate $\int xe^x\,dx$.

Solution   According to the list below,

$$u \xleftarrow{\quad\quad} \underset{\text{easier to differentiate}}{\ln x \quad (\sin^{-1} x, \tan^{-1} x) \quad (x^n, 1)} \quad \underset{\text{easier to integrate}}{(\sin x, \cos x) \quad e^x} \xrightarrow{\quad\quad} dv$$

x is closer to the left side than e^x. Let $u = x$ and $dv = e^x\,dx$. So, $\int xe^x\,dx = \int u\,dv$. Then

$$u = x \qquad\qquad v = e^x$$
$$du = dx \qquad\qquad dv = e^x\,dx$$

Thus,

$$\int u\,dv = uv - \int v\,du$$
$$= xe^x - \int e^x\,dx$$
$$= xe^x - e^x + C$$

Example 2 Evaluating the integral using integration by parts

Evaluate $\int \ln x\,dx$.

Solution $\int \ln x\,dx = \int \ln x \cdot 1\,dx$. According to the list below,

$$u \longleftarrow \underset{\text{easier to differentiate}}{\ln x \quad (\sin^{-1} x, \tan^{-1} x) \quad (x^n, 1) \quad (\sin x, \cos x)} \quad \underset{\text{easier to integrate}}{e^x} \longrightarrow dv$$

$\ln x$ is closer to the left side than 1. Let $u = \ln x$ and $dv = 1\,dx$. So, $\int \ln x\,dx = \int u\,dv$. Then

$$u = \ln x \qquad\qquad v = x$$
$$du = \frac{1}{x}\,dx \qquad\qquad dv = dx$$

Thus,

MR. RHEE'S BRILLIANT MATH SERIES   BC ONLY   AP CAL LESSON 33

$$\int u\,dv = uv - \int v\,du$$
$$= x\ln x - \int x \cdot \frac{1}{x}\,dx$$
$$= x\ln x - \int 1\,dx$$
$$= x\ln x - x + C$$

Evaluating a Definite Integral using Integration by Parts

$$\int_a^b u\,dv = uv\Big]_a^b - \int_a^b v\,du$$

Example 3 Evaluating the definite integral using integration by parts

Evaluate $\displaystyle\int_1^e \ln x\,dx$.

Solution As you obtained in Example 1, $\displaystyle\int \ln x\,dx = x\ln x - \int 1\,dx$. Thus,

$$\int_1^e \ln x\,dx = x\ln x\Big]_1^e - \int_1^e 1\,dx$$
$$= x\ln x\Big]_1^e - x\Big]_1^e$$
$$= e - (e - 1)$$
$$= 1$$

MR. RHEE'S BRILLIANT MATH SERIES — BC ONLY — AP CAL LESSON 33

EXERCISES

1. Evaluate $\int x \ln x \, dx$.

2. Evaluate $\int x^2 e^x \, dx$.

3. Evaluate $\int (\ln x)^2 \, dx$.

4. Evaluate $\int x \cos 3x \, dx$.

5. Evaluate $\int_0^1 \tan^{-1} x \, dx$.

MR. RHEE'S BRILLIANT MATH SERIES   BC ONLY   AP CAL LESSON 33

Answers

1. $\frac{1}{2}x^2 \ln x - \frac{1}{4}x^2 + C$
2. $x^2 e^x - 2xe^x + 2e^x + C$
3. $x(\ln x)^2 - 2x \ln x + 2x + C$
4. $\frac{1}{3}x \sin 3x + \frac{1}{9} \cos 3x + C$
5. $\frac{\pi}{4} - \frac{1}{2} \ln 2$

MR. RHEE'S BRILLIANT MATH SERIES   BC ONLY   AP CAL LESSON 34

LESSON 34

Trigonometric Integrals

We use the Pythagorean identities, double-angle and half-angle formulas to integrate some trigonometric functions involving powers of sine and cosine and powers of tangent and secant.

Pythagorean Identities

$$\sin^2 x + \cos^2 x = 1 \qquad 1 + \tan^2 x = \sec^2 x \qquad 1 + \cot^2 x = \csc^2 x$$

Tip   The following variations of the Pythagorean identities are often used.

$$\sin^2 x = 1 - \cos^2 x, \qquad \cos^2 x = 1 - \sin^2 x$$

Double-Angle and Half-Angle Formulas

- Double-Angle Formulas

$$\sin 2\theta = 2 \sin \theta \cos \theta \qquad \cos 2\theta = \cos^2 \theta - \sin^2 \theta$$
$$= 2\cos^2 \theta - 1$$
$$= 1 - 2\sin^2 \theta$$

- Half-Angle Formulas

$$\sin^2 x = \frac{1}{2}(1 - \cos 2x) \qquad \cos^2 x = \frac{1}{2}(1 + \cos 2x)$$

Example 1   Evaluating the trigonometric integral

Evaluate $\displaystyle \int \sin^3 x \, dx$.

MR. RHEE'S BRILLIANT MATH SERIES — BC ONLY — AP CAL LESSON 34

Solution Separate one sine factor and rewrite the remaining expression in terms of cosine.

$$\int \sin^3 x \, dx = \int \sin^2 x \cdot \sin x \, dx \qquad \text{Use the identity: } \sin^2 x = 1 - \cos^2 x$$

$$= \int (1 - \cos^2 x) \sin x \, dx$$

Then use the U-Substitution Rule. Let $u = \cos x$. $du = -\sin x \, dx$ gives $-du = \sin x \, dx$. Thus,

$$-\int (1 - u^2) \, du = -\left(u - \frac{1}{3}u^3\right) + C = -\cos x + \frac{1}{3}\cos^3 x + C$$

Evaluating $\int \sin^m x \cos^n x \, dx$

(a) When the power of sine is odd ($m = $ odd): separate one sine factor and use $\sin^2 x = 1 - \cos^2 x$ to rewrite the remaining expression in terms of cosine. For instance,

$$\int \sin^3 x \cos^2 x \, dx = \int \sin^2 x \cos^2 x \cdot \sin x \, dx$$

$$= \int (1 - \cos^2 x) \cos^2 x \cdot \sin x \, dx$$

(b) When the power of cosine is odd ($n = $ odd): separate one cosine factor and use $\cos^2 x = 1 - \sin^2 x$ to rewrite the remaining expression in terms of sine.

$$\int \sin^2 x \cos^3 x \, dx = \int \sin^2 x \cos^2 x \cdot \cos x \, dx$$

$$= \int \sin^2 x (1 - \sin^2 x) \cdot \sin x \, dx$$

(c) When the power of sine and cosine are even ($m = $ even and $n = $ even): use the half-angle formulas. For instance,

$$\int \sin^2 x \cos^2 x \, dx = \int \frac{1}{2}(1 - \cos 2x) \cdot \frac{1}{2}(1 + \cos 2x) \, dx$$

Example 2 Evaluating the trigonometric integral

Evaluate $\int \cos^5 x \sin^2 x \, dx$.

Solution Since the power of cosine is 5, separate one $\cos x$ factor and use $\cos^2 x = 1 - \sin^2 x$ to rewrite the remaining expression in terms of sine.

$$\int \cos^5 x \sin^2 x \, dx = \int \cos^4 x \sin^2 x \cdot \cos x \, dx$$
$$= \int (1 - \sin^2 x)^2 \sin^2 x \cdot \cos x \, dx$$

Let $u = \sin x$ and $du = \cos x \, dx$.

$$\int (1 - u^2)^2 u^2 \, du = \int (u^4 - 2u^2 + 1) u^2 \, du = \int u^6 - 2u^4 + u^2 \, du$$
$$= \frac{1}{7} u^7 - \frac{2}{5} u^5 + \frac{1}{3} u^3 + C$$
$$= \frac{1}{7} \sin^7 x - \frac{2}{5} \sin^5 x + \frac{1}{3} \sin^3 x + C$$

Evaluating $\int \tan^m x \sec^n x \, dx$

(a) When the power of tangent is odd ($m =$ odd): separate one $\sec x \tan x$ factor and use $\tan^2 x = \sec^2 x - 1$ to rewrite the remaining expression in terms of secant. For instance,

$$\int \tan^3 x \sec x \, dx = \int \tan^2 x \cdot \tan x \sec x \, dx$$
$$= \int (\sec^2 x - 1) \cdot \tan x \sec x \, dx$$

(b) When the power of secant is even ($n =$ even): separate one $\sec^2 x$ factor and use $\sec^2 x = 1 + \tan^2 x$ to rewrite the remaining expression in terms of tangent. For instance,

$$\int \tan^2 x \sec^4 x \, dx = \int \tan^2 x \sec^2 x \cdot \sec^2 x \, dx$$
$$= \int \tan^2 x (1 + \tan^2 x) \cdot \sec^2 x \, dx$$
$$= \int (\tan^2 x + \tan^4 x) \sec^2 x \, dx$$

MR. RHEE'S BRILLIANT MATH SERIES
BC ONLY   AP CAL LESSON 34

Example 3   Evaluating the trigonometric integral

Evaluate $\int \tan^3 x \sec^3 x \, dx$.

Solution
Since the power of tangent is odd, separate one $\sec x \tan x$ factor and use $\tan^2 x = \sec^2 x - 1$ to rewrite the remaining expression in terms of secant.

$$\int \tan^3 x \sec^3 x \, dx = \int \tan^2 x \sec^2 x \cdot \sec x \tan x \, dx$$
$$= \int (\sec^2 x - 1) \sec^2 x \sec x \tan x \, dx$$
$$= \int (\sec^4 x - \sec^2 x) \sec x \tan x \, dx$$

Let $u = \sec x$ and $du = \sec x \tan x \, dx$. Thus,

$$\int (u^4 - u^2) \, du = \frac{1}{5} u^5 - \frac{1}{3} u^3 + C$$
$$= \frac{1}{5} \sec^5 x - \frac{1}{3} \sec^3 x + C$$

EXERCISES

1. Evaluate $\int_0^{\frac{\pi}{2}} \sin^2 x \cos x \, dx$.

2. Evaluate $\int_0^{2\pi} \cos^2 x \, dx$.

3. Evaluate $\int \tan^2 x \, dx$.

4. Evaluate $\int \tan^3 x \, dx$.

5. Evaluate $\int \sin^4 x \, dx$.

MR. RHEE'S BRILLIANT MATH SERIES
BC ONLY — AP CAL LESSON 34

Answers

1. $\dfrac{1}{3}$
2. π
3. $\tan x - x + C$
4. $\dfrac{1}{2}\tan^2 x - \ln|\sec x| + C$
5. $\dfrac{1}{4}\left(\dfrac{3}{2}x - \sin 2x + \dfrac{1}{8}\sin 4x\right) + C$

LESSON 35

Integration By Partial Fractions

Partial Fraction Decomposition

If you subtract and simplify the rational functions below,

$$\frac{1}{x} - \frac{1}{x+1} = \frac{x+1-x}{x(x+1)} = \frac{1}{x(x+1)}$$

If you reverse the procedure,

$$\frac{1}{x(x+1)} = \frac{1}{x} - \frac{1}{x+1}$$

The procedure of writing a rational function as a sum of simpler fractions as shown above is called a **partial fraction decomposition**. The following guidelines enable us to find a partial fraction decomposition of a rational function.

Finding a partial fraction decomposition of a rational function

Factor the denominator of a rational function as far as possible. Then write the rational function as a sum of simpler fractions depending on types of factors the denominator has: a linear factor $(ax+b)$ or an irreducible quadratic factor (ax^2+bx+c). Simpler partial fraction are of the following forms.

$$\frac{A}{ax+b} \quad \text{or} \quad \frac{Ax+b}{ax^2+bx+c}$$

Below shows the four cases for the denominator.

Case 1: The denominator is a product of distinct linear factors. For instance,

$$\frac{1}{(x-1)(x+1)} = \frac{A}{x-1} + \frac{B}{x+1}$$

Case 2: The denominator is a product of linear factors, some of which are repeated. For instance,

$$\frac{1}{(x-1)(x+1)^3} = \frac{A}{x-1} + \frac{B}{x+1} + \frac{C}{(x+1)^2} + \frac{D}{(x+1)^3}$$

Case 3: The denominator is a product of a linear factor and an irreducible quadratic factor. For instance,

$$\frac{1}{(x-1)(x^2+1)} = \frac{A}{x-1} + \frac{Bx+C}{x^2+1}$$

MR. RHEE'S BRILLIANT MATH SERIES
BC ONLY — AP CAL LESSON 35

> **Case 4:** The denominator is a product of a linear factor and repeated irreducible quadratic factors. For instance,
>
> $$\frac{1}{(x-1)(x^2+1)^2} = \frac{A}{x-1} + \frac{Bx+C}{x^2+1} + \frac{Dx+E}{(x^2+1)^2}$$

Example 1 Finding a partial fraction decomposition

Find a partial fraction decomposition of $\dfrac{1}{x^2-1}$.

Solution Let's factor the denominator so that $x^2 - 1 = (x+1)(x-1)$. Since the denominator is a product of distinct linear factors, write $\dfrac{1}{x^2-1}$ as a sum of simpler fractions below.

$$\frac{1}{(x+1)(x-1)} = \frac{A}{x+1} + \frac{B}{x-1}$$

Multiplying both sides of the equation by $(x+1)(x-1)$ gives

$$1 = A(x-1) + B(x+1)$$

Consider the equation above as an identity equation, which is true for every value of x. Substituting $x = 1$ and $x = -1$ into the identity equation to find the coefficients of A and B,

$$\text{Substituting } x = 1: \quad 1 = 2B \quad \Longrightarrow \quad \therefore B = \frac{1}{2}$$

$$\text{Substituting } x = -1: \quad 1 = -2A \quad \Longrightarrow \quad \therefore A = -\frac{1}{2}$$

Thus, $\dfrac{1}{x^2-1} = -\dfrac{1}{2(x+1)} + \dfrac{1}{2(x-1)}$.

MR. RHEE'S BRILLIANT MATH SERIES   BC ONLY   AP CAL LESSON 35

Example 2   **Evaluating the integral by partial fractions**

Evaluate $\int \dfrac{1}{x^2-1}\,dx$.

Solution   As you obtained in Example 1, the partial fraction decomposition of $\dfrac{1}{x^2-1}$ is $-\dfrac{1}{2(x+1)} + \dfrac{1}{2(x-1)}$. Thus,

$$\begin{aligned}\int \dfrac{1}{x^2-1}\,dx &= \int \left(-\dfrac{1}{2(x+1)} + \dfrac{1}{2(x-1)}\right)dx \\ &= -\dfrac{1}{2}\int \dfrac{1}{x+1}\,dx + \dfrac{1}{2}\int \dfrac{1}{x-1}\,dx \\ &= -\dfrac{1}{2}\ln|x+1| + \dfrac{1}{2}\ln|x-1| + C\end{aligned}$$

MR. RHEE'S BRILLIANT MATH SERIES BC ONLY   AP CAL LESSON 35

EXERCISES

1. Evaluate $\int \dfrac{x^2+1}{x-1}\,dx$.

2. Evaluate $\int \dfrac{3x+2}{x^2+x}\,dx$.

3. Evaluate $\int \dfrac{6x+13}{x^2+5x+6}\,dx$.

MR. RHEE'S BRILLIANT MATH SERIES — BC ONLY — AP CAL LESSON 35

4. Evaluate $\displaystyle\int \frac{2x^2 - 5x - 8}{x^2 - 3x - 10}\, dx$.

5. Evaluate $\displaystyle\int \frac{2x^2 + x + 2}{x^3 + x}\, dx$.

MR. RHEE'S BRILLIANT MATH SERIES — BC ONLY — AP CAL LESSON 35

Answers

1. $\frac{1}{2}x^2 + x + 2\ln|x-1| + C$
2. $2\ln|x| + \ln|x+1| + C$
3. $\ln|x+2| + 5\ln|x+3| + C$
4. $2x - \frac{10}{7}\ln|x+2| + \frac{17}{7}\ln|x-5| + C$
5. $2\ln|x| + \tan^{-1}x + C$

MR. RHEE'S BRILLIANT MATH SERIES   BC ONLY   AP CAL LESSON 36

LESSON 36
Improper Integrals

> **Improper Integrals**
>
> An improper integral is a definite integral that has either or both limits infinite or an integrand has an infinite discontinuity in the domain. In order to evaluate the improper integral, we need to apply the limit to the definite integral. The improper integral is called **convergent** if the limit exists and **divergent** if the limit does not exist. Below summarizes types of improper integrals.
>
> 1. If $\int_a^t f(x)\,dx$ exists for every number $t \geq a$, then
>
> $$\int_a^\infty f(x)\,dx = \lim_{t \to \infty} \int_a^t f(x)\,dx$$
>
> 2. If $\int_t^b f(x)\,dx$ exists for every number $t \leq b$, then
>
> $$\int_{-\infty}^b f(x)\,dx = \lim_{t \to -\infty} \int_t^b f(x)\,dx$$
>
> 3. If $\int_{-\infty}^a f(x)\,dx$ and $\int_a^\infty f(x)\,dx$ are convergent, then
>
> $$\int_{-\infty}^\infty f(x)\,dx = \int_{-\infty}^a f(x)\,dx + \int_a^\infty f(x)\,dx$$
>
> 4. If f is continuous on $[a,b)$ and is discontinuous at b, then
>
> $$\int_a^b f(x)\,dx = \lim_{t \to b^-} \int_a^t f(x)\,dx$$
>
> 5. If f is continuous on $(a,b]$ and is discontinuous at a, then
>
> $$\int_a^b f(x)\,dx = \lim_{t \to a^+} \int_t^b f(x)\,dx$$
>
> 6. If f has a infinite discontinuity at c, where $a < c < b$, and both $\int_a^c f(x)\,dx$ and $\int_c^b f(x)\,dx$ are convergent, then
>
> $$\int_a^b f(x)\,dx = \int_a^c f(x)\,dx + \int_c^b f(x)\,dx$$

Example 1 Evaluating the improper integral

Evaluate $\displaystyle\int_1^\infty \frac{1}{x}\,dx$.

Solution

$$\begin{aligned}
\int_1^\infty \frac{1}{x}\,dx &= \lim_{t\to\infty} \int_1^t \frac{1}{x}\,dx \\
&= \lim_{t\to\infty} \ln|x|\Big]_1^t \\
&= \lim_{t\to\infty} (\ln t - \ln 1) \qquad \text{Since } t > 1,\ |t| = t \\
&= \ln(\infty) - 0 \\
&= \infty
\end{aligned}$$

Since the limit does not exists, the improper integral $\displaystyle\int_1^\infty \frac{1}{x}\,dx$ is divergent.

Example 2 Evaluating the improper integral

Evaluate $\displaystyle\int_1^\infty \frac{1}{x^2}\,dx$.

Solution

$$\begin{aligned}
\int_1^\infty \frac{1}{x^2}\,dx &= \lim_{t\to\infty} \int_1^t \frac{1}{x^2}\,dx \\
&= \lim_{t\to\infty} -\frac{1}{x}\Big]_1^t \\
&= -\lim_{t\to\infty}\left(\frac{1}{t} - 1\right) \\
&= -\left(\frac{1}{\infty} - 1\right) \\
&= 1
\end{aligned}$$

Since the limit exists, the improper integral $\displaystyle\int_1^\infty \frac{1}{x^2}\,dx$ is convergent.

Tip In general, $\displaystyle\int_1^\infty \frac{1}{x^p}\,dx$ is convergent if $p > 1$ and divergent if $p \leq 1$.

Example 3 Evaluating the improper integral

Evaluate $\displaystyle\int_{-\infty}^{\infty} \frac{1}{1+x^2}\,dx$.

Solution

$$\int_{-\infty}^{\infty} \frac{1}{1+x^2}\,dx = \int_{-\infty}^{a} \frac{1}{1+x^2}\,dx + \int_{a}^{\infty} \frac{1}{1+x^2}\,dx$$

We choose $a = 0$ for simplicity. Thus,

$$\begin{aligned}
\int_{-\infty}^{\infty} \frac{1}{1+x^2}\,dx &= \int_{-\infty}^{0} \frac{1}{1+x^2}\,dx + \int_{0}^{\infty} \frac{1}{1+x^2}\,dx \\
&= \lim_{t\to -\infty} \int_{t}^{0} \frac{1}{1+x^2}\,dx + \lim_{t\to \infty} \int_{0}^{t} \frac{1}{1+x^2}\,dx \\
&= \lim_{t\to -\infty} \tan^{-1} x \Big]_{t}^{0} + \lim_{t\to \infty} \tan^{-1} x \Big]_{0}^{t} \\
&= \lim_{t\to -\infty} (\tan^{-1} 0 - \tan^{-1} t) + \lim_{t\to \infty} (\tan^{-1} t - \tan^{-1} 0) \\
&= \left(\tan^{-1} 0 - \tan^{-1}(-\infty)\right) + \left(\tan^{-1}(\infty) - \tan^{-1} 0\right)
\end{aligned}$$

The figure below shows the graph of $y = \tan^{-1} x$.

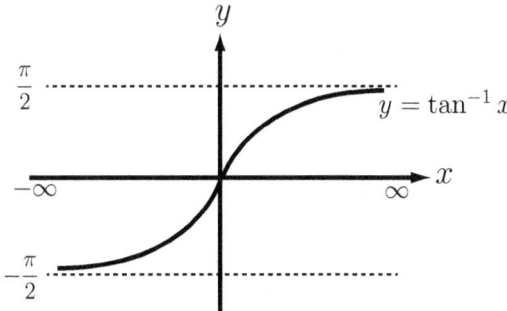

As $x \to \infty$, $f(x) \to \dfrac{\pi}{2}$. As $x \to -\infty$, $f(x) \to -\dfrac{\pi}{2}$. Thus,

$$\begin{aligned}
\left(\tan^{-1} 0 - \tan^{-1}(-\infty)\right) + \left(\tan^{-1}(\infty) - \tan^{-1} 0\right) &= 0 - \left(-\frac{\pi}{2}\right) + \frac{\pi}{2} - 0 \\
&= \frac{\pi}{2} + \frac{\pi}{2} \\
&= \pi
\end{aligned}$$

MR. RHEE'S BRILLIANT MATH SERIES BC ONLY AP CAL LESSON 36

EXERCISES

1. Evaluate $\displaystyle\int_0^\infty e^{-x}\,dx$.

2. Evaluate $\displaystyle\int_0^2 \frac{1}{x^2}\,dx$.

3. Evaluate $\displaystyle\int_0^1 \ln x\,dx$.

MR. RHEE'S BRILLIANT MATH SERIES — BC ONLY — AP CAL LESSON 36

4. Evaluate $\int_1^3 \dfrac{1}{x-2}\, dx$.

5. Evaluate $\int_{-\infty}^{\infty} x e^{-x^2}\, dx$.

MR. RHEE'S BRILLIANT MATH SERIES

BC ONLY AP CAL LESSON 36

Answers

1. 1
2. Divergent
3. −1
4. Divergent
5. 0

LESSON 37

Differential Equations

Euler's Method

Finding a explicit solution of a differential equation can be hard or impossible. Thus, we use a numerical approach: Euler's method. It enables us to find a numerical approximations to the solution of a differential equation despite the explicit solution. The Euler's method is defined by recursive formulas

$$x_n = x_{n-1} + h, \qquad y_n = y_{n-1} + h \left.\frac{dy}{dx}\right|_{x_{n-1}, y_{n-1}}$$

where h is called the step size, which represents the horizontal distance traveled.

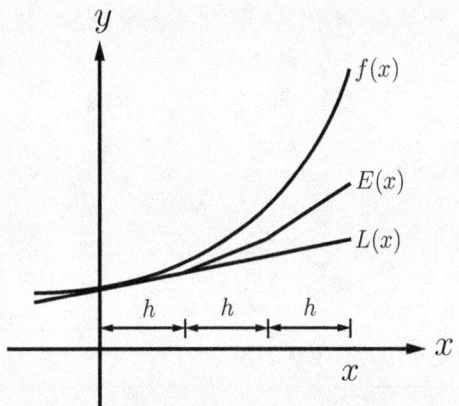

In the figure above, $L(x)$ is the linear approximation of $f(x)$. Euler's method, $E(x)$, starts at the initial value and proceed a step size along the tangent line and then change its direction according to the slope field. Thus, $E(x)$ gives a better approximation of $f(x)$ than $L(x)$. As we decrease the step size h, we will obtain better approximation to the exact solution.

MR. RHEE'S BRILLIANT MATH SERIES BC ONLY AP CAL LESSON 37

Example 1 Using Euler's method

Use Euler's method with step size $h = \dfrac{1}{3}$ to estimate the value of $y(1)$, where y is the solution of the initial-value problem $y' = y$, $y(0) = 1$.

Solution

$$x_0 = 0 \qquad y_0 = 1$$

$$x_1 = x_0 + h = \frac{1}{3} \qquad y_1 = y_0 + h \left.\frac{dy}{dx}\right|_{x_0, y_0} = 1 + \frac{1}{3}\left.\frac{dy}{dx}\right|_{0,1} = 1 + \frac{1}{3} = \frac{4}{3}$$

$$x_2 = x_1 + h = \frac{2}{3} \qquad y_2 = y_1 + h \left.\frac{dy}{dx}\right|_{x_1, y_1} = \frac{4}{3} + \frac{1}{3}\left.\frac{dy}{dx}\right|_{\frac{1}{3},\frac{4}{3}} = \frac{4}{3} + \frac{1}{3} \cdot \frac{4}{3} = \frac{16}{9}$$

$$x_3 = x_2 + h = 1 \qquad y_3 = y_2 + h \left.\frac{dy}{dx}\right|_{x_2, y_2} = \frac{16}{9} + \frac{1}{3}\left.\frac{dy}{dx}\right|_{\frac{2}{3},\frac{16}{9}} = \frac{16}{9} + \frac{1}{3} \cdot \frac{16}{9} = 2.370$$

Therefore, the value of $y(1)$ using the Euler's method with step size $h = \dfrac{1}{3}$ is 2.370.

Solving Separable Equations by Parts

A differential equation that can be solved explicitly or implicitly is called **separable equation**. A separable equation is a first-order differential equation in which the variables can be separated so that one variable is one side of the equation and the other variable on the other side of the equation. The following guidelines will help you solve a separable equation.

Guidelines for solving a separable equation

1. Separable the variables: group $g(y)$ and dy on the left side, and $f(x)$ and dx on the right side.

2. Integrate both sides of the equation.

3. Write the general solution explicitly or implicitly.

MR. RHEE'S BRILLIANT MATH SERIES BC ONLY AP CAL LESSON 37

Example 2 Solving separable equation by parts

Solve the differential equation $\dfrac{dy}{dx} = \dfrac{xe^x}{y}$.

Solution
Separate the variables first.

$$\frac{dy}{dx} = \frac{xe^x}{y}$$
$$y\, dy = xe^x\, dx \qquad \text{Integrate both sides}$$
$$\int y\, dy = \int xe^x\, dx$$

Next, evaluate $xe^x\, dx$. According to the list below,

$$u \xleftarrow{\underset{\text{easier to differentiate}}{}} \ln x \quad (\sin^{-1} x, \tan^{-1} x) \quad (x^n, 1) \quad (\sin x, \cos x) \quad e^x \xrightarrow{\underset{\text{easier to integrate}}{}} dv$$

x is closer to the left side than e^x. Let $u = x$ and $dv = e^x\, dx$. So, $\int xe^x\, dx = \int u\, dv$. Then

$$u = x \qquad\qquad v = e^x$$
$$du = dx \qquad\qquad dv = e^x\, dx$$

Thus,

$$\int u\, dv = uv - \int v\, du$$
$$= xe^x - \int e^x\, dx$$
$$= xe^x - e^x + C$$

Therefore,

$$\int y\, dy = \int xe^x\, dx$$
$$\frac{1}{2}y^2 = xe^x - e^x + C$$

MR. RHEE'S BRILLIANT MATH SERIES — BC ONLY — AP CAL LESSON 37

Logistic Differential Equation

Due to limited resources, a population increases exponentially in its early stages but levels off and approaches its **carrying capacity**, the maximum sustainable population. A differential equation that well describes the population growth behavior is known as the **logistic differential equation** shown below.

$$\frac{dP}{dt} = kP\left(1 - \frac{P}{C}\right)$$

where C is the carrying capacity. The general solution of the logistic differential equation is given by

$$P(t) = \frac{C}{1 + \frac{C - P_0}{P_0} e^{-kt}}$$

where P_0 is the initial population. The figure below shows the graph of the logistic differential equation.

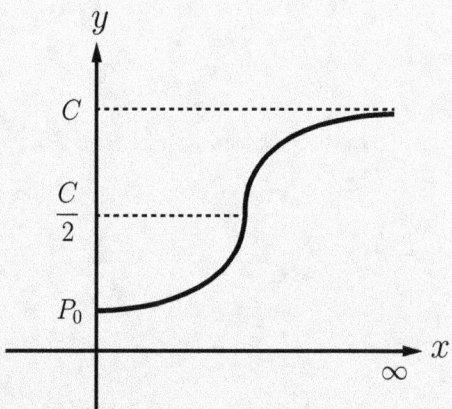

Tip

1. For the logistic differential equation, the growth rate of a population is fastest when the population reaches half its carrying capacity.

2. As $t \to \infty$, the population reaches its carrying capacity. In other words, $\lim\limits_{t \to \infty} \dfrac{C}{1 + \frac{C - P_0}{P_0} e^{-kt}} = C$.

Example 3 Solving the logistic differential equation

Suppose that a population grows according to the logistic differential equation $\dfrac{dP}{dt} = 0.02P - 0.0001P^2$. When $t = 0$, the population is 20.

(a) Find the carrying capacity.

(b) Find the solution of the logistic differential equation.

Solution

(a) Rewrite $\dfrac{dP}{dt} = 0.02P - 0.0001P^2$ as $\dfrac{dP}{dt} = 0.02P\left(1 - \dfrac{P}{200}\right)$. Thus, the carrying capacity is 200.

(b) The solution of the differential equation is

$$P(t) = \dfrac{C}{1 + \dfrac{C - P_0}{P_0} e^{-kt}}$$

$$= \dfrac{200}{1 + \dfrac{200 - 20}{20} e^{-0.02t}}$$

$$= \dfrac{200}{1 + 9 e^{-0.02t}}$$

MR. RHEE'S BRILLIANT MATH SERIES
BC ONLY   AP CAL LESSON 37

EXERCISES

1. Use Euler's method with step size 0.2 to estimate the value of $y(0.6)$, where y is the solution of the initial-value problem $y' = x + y$, $y(0) = 2$.

2. Solve the differential equation $\dfrac{dy}{dx} = \dfrac{e^{2x}}{\ln y}$.

3. Find the solution of the differential equation $e^{-t}\dfrac{dx}{dt} = \dfrac{t}{x}$ that satisfies the given initial condition $x(0) = 1$.

4. Suppose that a population grows according to the logistic differential equation with carrying capacity $10,000$ and $k = 0.02$.

   (a) Write the logistic differential equation.

   (b) If the initial population is $2,000$, write a formula for the population, $P(t)$, after t years.

   (c) Find the population after 10 years.

   (d) Find the time at which the growth rate of the population is fastest.

   (e) Find $\lim\limits_{t \to \infty} P(t)$.

MR. RHEE'S BRILLIANT MATH SERIES BC ONLY   AP CAL LESSON 37

Answers

1. $y(0.6) = 3.584$
2. $y \ln y - y = \frac{1}{2} e^{2x} + C$
3. $\frac{1}{2} x^2 = te^t - e^t + \frac{3}{2}$
4(a). $\frac{dP}{dt} = 0.02 P \left(1 - \frac{P}{10,000}\right)$
4(b). $P(t) = \dfrac{10,000}{1 + 4e^{-0.02t}}$
4(c). $P(10) = 2339.22$
4(d). 69.31
4(e). $\lim_{t \to \infty} P(t) = 10,000$

LESSON 38

Derivative, Arc Length, And Area With Polar Coordinates

Polar Coordinates

In rectangular coordinates shown in Figure 1, a point is determined by (x, y), where x, and y represent where the point is x units horizontally and y units vertically from the origin, respectively.

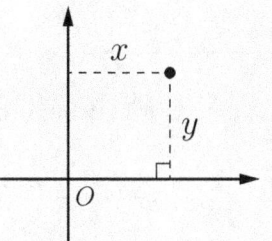
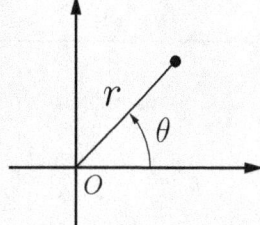

Fig. 1: Rectangular Coordinates   Fig. 2: Polar Coordinates

Whereas, in polar coordinates shown in Figure 2, a point is determined by (r, θ), where r represents the distance between the point and the origin, and θ represents the counterclockwise angle formed by the positive x-axis and the terminal side.

Converting from Polar Coordinates to Rectangular Coordinates

If the polar coordinates of a point is (r, θ), the rectangular coordinates of the point (x, y) are given by

$$x = r\cos\theta, \qquad y = r\sin\theta$$

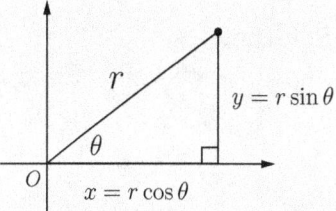

For instance, if the polar coordinates of a point is $(2, \frac{3\pi}{4})$, the rectangular coordinates of the point are as follows:

$$x = r\cos\theta \quad \Longrightarrow \quad x = 2\cos\frac{3\pi}{4} = 2\left(-\frac{\sqrt{2}}{2}\right) = -\sqrt{2}$$

$$y = r\sin\theta \quad \Longrightarrow \quad y = 2\sin\frac{3\pi}{4} = 2\left(\frac{\sqrt{2}}{2}\right) = \sqrt{2}$$

Thus, the rectangular coordinates of the point is $(-\sqrt{2}, \sqrt{2})$.

MR. RHEE'S BRILLIANT MATH SERIES
BC ONLY — AP CAL LESSON 38

Derivative of a Polar Equation

Suppose a polar curve is defined by a polar equation $r = f(\theta)$. Then

$$x = r\cos\theta = f(\theta)\cos\theta, \qquad y = r\sin\theta = f(\theta)\sin\theta$$

Using the method for finding the slope of the tangent line to a parametric curve, the slope of the tangent line to the polar curve is given by

$$\frac{dy}{dx} = \frac{\frac{dy}{d\theta}}{\frac{dx}{d\theta}} = \frac{\frac{dr}{d\theta}\sin\theta + r\cos\theta}{\frac{dr}{d\theta}\cos\theta - r\sin\theta}$$

Tip

1. The polar curve defined by polar equation has a horizontal tangent when $\dfrac{dy}{d\theta} = 0$ and it has a vertical tangent when $\dfrac{dx}{d\theta} = 0$

2. $\dfrac{dy}{dx} = \dfrac{dy}{d\theta} \Big/ \dfrac{dx}{d\theta}$ enables us to find the slope the tangent line to the polar curve without having to eliminate the parameter θ.

Example 1 Finding the slope of the tangent to the polar curve

Find the slope of the tangent line to the polar curve $r = 2\cos\theta$ at $\theta = \dfrac{\pi}{3}$.

Solution

$$y = r\sin\theta = 2\cos\theta\sin\theta = \sin 2\theta$$
$$x = r\cos\theta = 2\cos^2\theta$$

Thus,

$$\frac{dy}{dx} = \frac{\frac{dy}{d\theta}}{\frac{dx}{d\theta}} = \frac{2\cos 2\theta}{-4\cos\theta\sin\theta}$$

$$\left.\frac{dy}{dx}\right|_{\theta=\frac{\pi}{3}} = \frac{2\cos\frac{2\pi}{3}}{-4\cos\frac{\pi}{3}\sin\frac{\pi}{3}} = \frac{\sqrt{3}}{3}$$

Therefore, the slope of the tangent line to the polar curve $r = 2\cos\theta$ at $\theta = \dfrac{\pi}{3}$ is $\dfrac{\sqrt{3}}{3}$.

MR. RHEE'S BRILLIANT MATH SERIES BC ONLY AP CAL LESSON 38

Arc Length of a Polar Equation

If a curve is defined by a polar equation $r = f(\theta)$, where $\alpha \leq \theta \leq \beta$, then the arc length L of the curve is given by

$$L = \int_\alpha^\beta \sqrt{r^2 + \left(\frac{dr}{d\theta}\right)^2}\, d\theta$$

Tip

1. Recall that the arc length L of the curve $y = f(x)$ for $a \leq x \leq b$ is

$$L = \int_a^b \sqrt{1 + \left(\frac{dy}{dx}\right)^2}\, dx$$

2. Recall that the arc length L of the curve given by parametric equations $x = f(t)$ and $y = g(t)$ for $\alpha \leq t \leq \beta$ is

$$L = \int_\alpha^\beta \sqrt{\left(\frac{dx}{dt}\right)^2 + \left(\frac{dy}{dt}\right)^2}\, dt$$

Example 2 Finding the arc length of the polar curve

Find the arc length of the curve defined by $r = 2\cos\theta$ for $0 \leq \theta \leq \pi$.

Solution $\dfrac{dr}{d\theta} = -2\sin\theta$ gives $\left(\dfrac{dr}{d\theta}\right)^2 = 4\sin^2\theta$. Thus,

$$\begin{aligned}
L &= \int_\alpha^\beta \sqrt{r^2 + \left(\frac{dr}{d\theta}\right)^2}\, d\theta \\
&= \int_0^\pi \sqrt{4\cos^2\theta + 4\sin^2\theta}\, d\theta \\
&= \int_0^\pi \sqrt{4(\cos^2\theta + \sin^2\theta)}\, d\theta \qquad \text{Since } \cos^2\theta + \sin^2\theta = 1 \\
&= \int_0^\pi 2\, d\theta \\
&= 2\theta \Big|_0^\pi \\
&= 2\pi
\end{aligned}$$

Area between two Polar Equations

- Suppose the region R is bounded by the polar curve $r = f(\theta)$ and the two rays $\theta = a$ and $\theta = b$ as shown in the figure below.

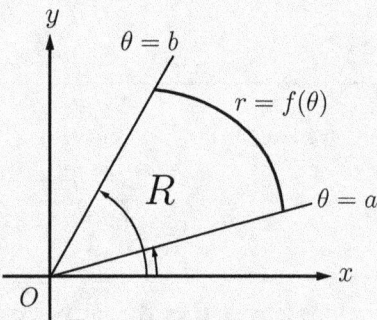

Then the area A of the region R is

$$A = \frac{1}{2} \int_a^b [f(\theta)]^2 \, d\theta$$

- Suppose the region R is bounded by the polar curves $r = f(\theta)$, $r = g(\theta)$, and the two rays $\theta = a$ and $\theta = b$ as shown in the figure below.

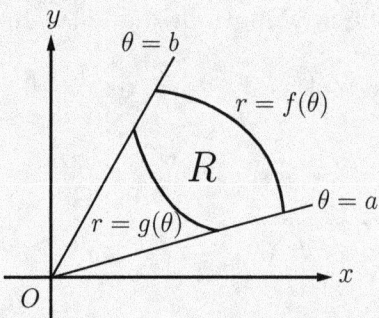

Then the area A of the region R is

$$A = \frac{1}{2} \int_a^b [f(\theta)]^2 - [g(\theta)]^2 \, d\theta$$

Example 3 Finding the area of the region bounded by the polar curve

Find the area bounded by one loop of the polar curve $r = \cos 2\theta$.

Solution The figure below shows the graph of the polar curve $r = \cos 2\theta$.

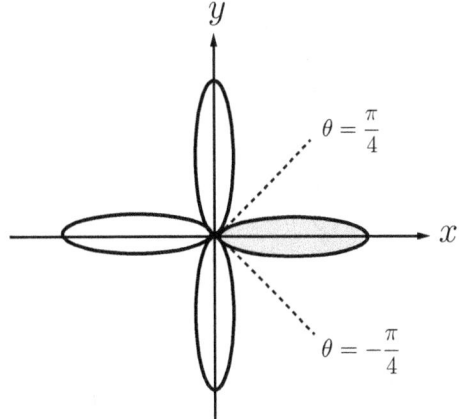

The one of the polar curve is bounded by $\theta = -\dfrac{\pi}{4}$ and $\theta = \dfrac{\pi}{4}$. Thus, the area A bounded by one loop of the polar curve $r = \cos 2\theta$ is

$$A = \frac{1}{2}\int_a^b [f(\theta)]^2\, d\theta = \frac{1}{2}\int_{-\frac{\pi}{4}}^{\frac{\pi}{4}} \cos^2 2\theta\, d\theta$$

$$= \frac{1}{2}\int_{-\frac{\pi}{4}}^{\frac{\pi}{4}} \frac{1}{2}(1 + \cos 4\theta)\, d\theta$$

Since the graph is symmetric about the x-axis, $\dfrac{1}{2}\int_{-\frac{\pi}{4}}^{\frac{\pi}{4}} \dfrac{1}{2}(1 + \cos 4\theta)\, d\theta = \dfrac{1}{2} \cdot 2 \int_0^{\frac{\pi}{4}} \dfrac{1}{2}(1 + \cos 4\theta)\, d\theta$. Thus,

$$\frac{1}{2}\int_{-\frac{\pi}{4}}^{\frac{\pi}{4}} \frac{1}{2}(1 + \cos 4\theta)\, d\theta = \int_0^{\frac{\pi}{4}} \frac{1}{2}(1 + \cos 4\theta)\, d\theta$$

$$= \frac{1}{2}\left[\theta + \frac{1}{4}\sin 4\theta\right]_0^{\frac{\pi}{4}}$$

$$= \frac{1}{2} \cdot \frac{\pi}{4}$$

$$= \frac{\pi}{8}$$

Therefore, the area bounded by one loop of the polar curve $r = \cos 2\theta$ is $\dfrac{\pi}{8}$.

MR. RHEE'S BRILLIANT MATH SERIES BC ONLY AP CAL LESSON 38

EXERCISES

1. Consider the polar curve $r = 1 + \cos\theta$ for $0 \leq \theta < 2\pi$.

    (a) Find $\dfrac{dy}{dx}$.

    (b) Find the points at which the polar curve has a vertical tangent.

2. Find an equation of the tangent line to the polar curve $r = \sin\theta + \cos\theta$ at $\theta = \dfrac{\pi}{4}$.

3. Set up, but do not evaluate, an integral that represents the arc length of the polar curve $r = 1 + \sin\theta$ for $0 \leq \theta \leq \pi$.

4. Find the area of the region bounded by $r = 4\sin\theta$ and $r = 2$.

5. Find the area of the region that lies inside the curve $r = 3\sin\theta$ and outside the curve $r = 1 + \sin\theta$.

MR. RHEE'S BRILLIANT MATH SERIES   BC ONLY   AP CAL LESSON 38

Answers

1(a). $\dfrac{dy}{dx} = \dfrac{\cos\theta + \cos^2\theta - \sin^2\theta}{-\sin\theta(1 + 2\cos\theta)}$

1(b). $\theta = 0,\ \pi,\ \dfrac{2\pi}{3},\ \dfrac{4\pi}{3}$

2. $y - 1 = -(x - 1)$

3. $L = \displaystyle\int_0^\pi \sqrt{(1 + \sin\theta)^2 + \cos^2\theta}\, d\theta$

4. $\dfrac{4\pi}{3} + 2\sqrt{3}$

5. π

LESSON 39

Sequences

Sequences

A sequence is a list of numbers in order. The numbers in the list are called **terms** of the sequence and are denoted with subscripted letters: a_1 for the first term, a_2 for the second term, a_n for the nth term.

For every positive integer n, there is a corresponding number a_n. So, a sequence can be defined as a function whose domain is the set of positive integers.

The following sequence $\left\{\dfrac{1}{2}, \dfrac{2}{3}, \dfrac{3}{4}, \dfrac{4}{5}, \cdots\right\}$ can also be defined by

$$\left\{\frac{n}{n+1}\right\}_{n=1}^{\infty} \quad \text{or} \quad a_n = \frac{n}{n+1}$$

The Limit of a Sequence

A sequence a_n has the limit L if $\lim\limits_{n\to\infty} a_n = L$. If $\lim\limits_{n\to\infty} a_n$ exits, we say that the sequence converges. Otherwise, the sequence diverges.

Limit properties for Sequences

Taking the limit of a sequence is identical to taking the limit of a function. Suppose $\{a_n\}$ and $\{b_n\}$ are convergent sequences and c is a constant. Then the limit properties for sequences are as follows:

1. $\lim\limits_{n\to\infty}(a_n \pm b_n) = \lim\limits_{n\to\infty} a_n \pm \lim\limits_{n\to\infty} b_n$

2. $\lim\limits_{n\to\infty} ca_n = c \lim\limits_{n\to\infty} a_n$

3. $\lim\limits_{n\to\infty}(a_n \cdot b_n) = \lim\limits_{n\to\infty} a_n \cdot \lim\limits_{n\to\infty} b_n$

4. $\lim\limits_{n\to\infty} \dfrac{a_n}{b_n} = \dfrac{\lim\limits_{n\to\infty} a_n}{\lim\limits_{n\to\infty} b_n}$, if $\lim\limits_{n\to\infty} b_n \ne 0$

5. $\lim\limits_{n\to\infty} c = c$

Theorems for Sequences

Below summarizes useful theorems for sequences.

1. Squeeze theorem for sequences:
   If $a_n \leq b_n \leq c_n$ and $\lim_{n \to \infty} a_n = \lim_{n \to \infty} c_n = L$, then $\lim_{n \to \infty} b_n = L$.

2. If $\lim_{n \to \infty} |a_n| = 0$, then $\lim_{n \to \infty} a_n = 0$.

3. If $\lim_{x \to \infty} f(x) = L$ and $f(n) = a_n$ when n is an integer, then $\lim_{n \to \infty} a_n = L$.

4. The sequence $\{r^n\}$ is convergent if $-1 < r \leq 1$ and divergent for all other values of r.

$$\lim_{n \to \infty} r^n = \begin{cases} 0, & \text{if } -1 < r < 1 \\ 1, & \text{if } r = 1 \end{cases}$$

5. A sequence $\{a_n\}$ is called **increasing** if $a_n \leq a_{n+1}$ for all $n \geq 1$. Whereas, it is called **decreasing** if $a_n \geq a_{n+1}$ for all $n \geq 1$. It is called **monotonic** if it is either increasing or decreasing.

   A sequence $\{a_n\}$ is **bounded above** if there is a number M such that $a_n \leq M$ for all $n \geq 1$ as shown in Figure 1.

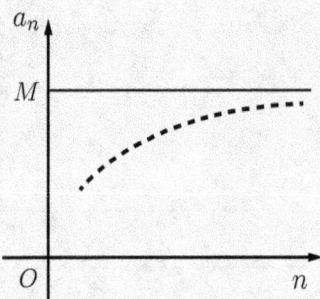
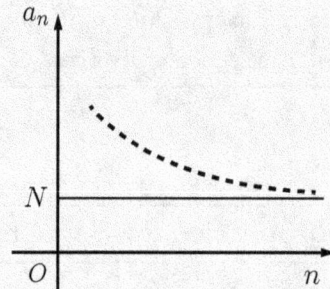

Figure 1:Bounded above   Figure 2:Bounded Below

Whereas, it is **bounded below** if there is a number N such that $N \leq a_n$ for all $n \geq 1$ as shown in Figure 2. If a sequence is either bounded above or bounded below, then the sequence is a bounded sequence.

Monotonic Sequence Theorem:
Every bounded, monotonic sequence is convergent

MR. RHEE'S BRILLIANT MATH SERIES BC ONLY   AP CAL LESSON 39

Example 1 Determining whether the sequence converges or diverges

Determine whether the sequence $\left\{\dfrac{n^2+n-1}{2n^2-3n+4}\right\}_1^\infty$ converges or diverges.

Solution In order to determine whether the sequence converges or diverges, take the limit of the function at infinity. Notice that $\lim\limits_{n\to\infty}\dfrac{n^2+n-1}{2n^2-3n+4}$ is identical to finding the horizontal asymptote of the rational function $\dfrac{n^2+n-1}{2n^2-3n+4}$. Since the degree of the numerator and the denominator of the rational function are the same, the horizontal asymptote of the rational function is the ratio of leading coefficients. Thus,

$$\lim_{n\to\infty}\frac{n^2+n-1}{2n^2-3n+4}=\frac{1}{2}$$

Therefore, the sequence $\left\{\dfrac{n^2+n-1}{2n^2-3n+4}\right\}_1^\infty$ converges.

Example 2 Determining whether the sequence converges or diverges

Determine whether the sequence $\left\{\dfrac{e^{2n}}{n}\right\}_1^\infty$ converges or diverges.

Solution Let $f(x)=\dfrac{e^{2x}}{x}$ so that $f(n)=a_n$. In order to determine whether the sequence converges or diverges, evaluate $\lim\limits_{x\to\infty}\dfrac{e^{2x}}{x}$. Plugging-in $x=\infty$ into the numerator and the denominator, we get an indeterminate form of $\dfrac{\infty}{\infty}$. Use the L'Hospital's Rule to find the limit.

$$\lim_{x\to\infty}\frac{e^{2x}}{x}=\lim_{x\to\infty}\frac{(e^{2x})'}{(x)'}$$
$$=\lim_{x\to\infty}\frac{2e^{2x}}{1}$$
$$=\infty$$

Therefore, the sequence $\left\{\dfrac{e^{2n}}{n}\right\}_1^\infty$ diverges.

MR. RHEE'S BRILLIANT MATH SERIES
BC ONLY   AP CAL LESSON 39

EXERCISES

For questions 1-4, determine whether the following sequences converge or diverge.

1. $\left\{\cos \pi n\right\}_1^\infty$

2. $\left\{\tan^{-1} n\right\}_1^\infty$

3. $a_n = \dfrac{3^n}{4^{n+1}}$

4. $\left\{\dfrac{n}{\ln n}\right\}_1^\infty$

5. Determine whether the sequence $a_n = \dfrac{n}{n^2+1}$ is increasing or decreasing or not monotonic.

MR. RHEE'S BRILLIANT MATH SERIES BC ONLY   AP CAL LESSON 39

Answers

1. Diverges
2. Converges
3. Converges
4. Diverges
5. a_n is decreasing

MR. RHEE'S BRILLIANT MATH SERIES   BC ONLY   AP CAL LESSON 40

LESSON 40

Convergence And Divergence Of Series, Part I

Definition of Convergent and Divergent Series

The nth partial sum of the series $\sum_{n=1}^{\infty} a_n$ is given by $S_n = a_1 + a_2 + a_3 + \cdots + a_n$.

$$S_1 = a_1$$
$$S_2 = a_1 + a_2$$
$$S_3 = a_1 + a_2 + a_3$$
$$\vdots \qquad \vdots$$
$$S_n = a_1 + a_2 + a_3 + \cdots + a_n = \sum_{i=1}^{n} a_i$$

If the sequence of these partial sum $\{S_n\}$ shown above converges to S, then the $\sum_{n=1}^{\infty} a_n$ is called **convergent**. Otherwise the series is called **divergent**. The real number S is called the **sum** of the series.

Determining whether a series converges or diverges is not an easy task. You will learn about **ten convergence tests** for the series. With the following convergence tests, you will be able to evaluate almost any series for its convergence.

Test 1: Test for Divergence

If $\lim_{n \to \infty} a_n \neq 0$, then the series $\sum_{n=1}^{\infty} a_n$ is divergent.

[Tip]
1. The contrapositive of the Test for Divergence is also true.

   $\sum_{n=1}^{\infty} a_n$ is convergent, then $\lim_{n \to \infty} a_n = 0$.

2. Note that the following statement is **NOT** true in general.

   If $\lim_{n \to \infty} a_n = 0$, then $\sum_{n=1}^{\infty} a_n$ is convergent.

MR. RHEE'S BRILLIANT MATH SERIES   BC ONLY   AP CAL LESSON 40

Test 2: Geometric Series Test

Given the geometric series $\sum_{n=1}^{\infty} ar^{n-1} = a + ar + ar^2 + \cdots$, it is convergent if $|r| < 1$, and its sum is $\dfrac{a}{1-r}$. Otherwise, the geometric series is divergent if $|r| \geq 1$.

$$\sum_{n=1}^{\infty} ar^{n-1} = \begin{cases} \dfrac{a}{1-r}, & |r| < 1 \\ \infty, & |r| \geq 1 \end{cases}$$

Test 3: P-Series Test

The p-series $\sum_{n=1}^{\infty} \dfrac{1}{n^p}$ is convergent if $p > 1$ and divergent if $p \leq 1$.

(Tip) The p-series with $p = 1$ is called the **harmonic series** and is divergent.

$$\sum_{n=1}^{\infty} \frac{1}{n} = 1 + \frac{1}{2} + \frac{1}{3} + \frac{1}{4} + \cdots = \infty$$

MR. RHEE'S BRILLIANT MATH SERIES
BC ONLY — AP CAL LESSON 40

Test 4: Integral Test

Suppose f is a continuous, positive, decreasing function on $[1, \infty]$ and let $a_n = f(n)$. Then $\int_1^\infty f(x)\,dx$ and $\sum_{n=1}^\infty a_n$ either both converge or both diverge. In other words,

1. If $\int_1^\infty f(x)\,dx$ converges, then $\sum_{n=1}^\infty a_n$ converges.

2. If $\int_1^\infty f(x)\,dx$ diverges, then $\sum_{n=1}^\infty a_n$ diverges.

Tip

1. It is not necessary to start the series or integral at $n = 1$. Thus,

$$\text{For } \sum_{n=3}^\infty a_n \quad \Longrightarrow \quad \text{Use } \int_3^\infty f(x)\,dx$$

2. It is not necessary that f is always decreasing. It is ok to use the integral test as long as f is decreasing for large value of x.

Remainder Estimate for the Integral Test

If $\sum_{n=1}^\infty a_n$ converges by the Integral Test and $R_n = S - S_n$, then

$$\int_{n+1}^\infty f(x)\,dx \le R_n \le \int_n^\infty f(x)\,dx$$

where R_n is the remainder (error) involved in approximating the exact sum S of the infinite series by the partial sum S_n.

MR. RHEE'S BRILLIANT MATH SERIES
BC ONLY — AP CAL LESSON 40

Test 5: Alternating Series Test

If the alternating series

$$\sum_{n=1}^{\infty}(-1)^n a_n = -a_1 + a_2 - a_3 + \cdots \quad \text{or} \quad \sum_{n=1}^{\infty}(-1)^{n-1} a_n = a_1 - a_2 + a_3 + \cdots$$

satisfies the following two conditions

1. $a_{n+1} \leq a_n$ for all n
2. $\lim_{n \to \infty} a_n = 0$.

then the series converges.

Tip: When considering the two conditions, ignore $(-1)^n$ or $(-1)^{n-1}$ in the series.

Alternating Series Estimation Theorem

If an alternating series $\sum_{n=1}^{\infty}(-1)^n a_n$ satisfies the two conditions

1. $a_{n+1} \leq a_n$ for all n
2. $\lim_{n \to \infty} a_n = 0$.

and converges to S, then the remainder (error) $R_n = S - S_n$ is

$$|R_n| = |S - S_n| \leq a_{n+1}$$

Tip: Later, alternating series estimation theorem will be used again to find the error involved in approximation by the Taylor or Maclaurin series.

MR. RHEE'S BRILLIANT MATH SERIES   BC ONLY   AP CAL LESSON 40

Example 1   Determining whether the series is convergent or divergent

Determine whether the series $\sum_{n=1}^{\infty} \dfrac{n+1}{\sqrt{2n^2+n+3}}$ is convergent or divergent.

Solution

$$\lim_{n\to\infty} \frac{n+1}{\sqrt{2n^2+n+3}} = \lim_{n\to\infty} \frac{n+1}{\sqrt{2n^2+n+3}} \cdot \frac{\frac{1}{n}}{\frac{1}{n}}$$

$$= \lim_{n\to\infty} \frac{1+\frac{1}{n}}{\sqrt{\frac{2n^2+n+3}{n^2}}}$$

$$= \lim_{n\to\infty} \frac{1+\frac{1}{n}}{\sqrt{2+\frac{1}{n}+\frac{3}{n^2}}}$$

$$= \frac{1}{\sqrt{2}}$$

Since $\lim_{n\to\infty} \dfrac{n+1}{\sqrt{2n^2+n+3}} \neq 0$, the series $\sum_{n=1}^{\infty} \dfrac{n+1}{\sqrt{2n^2+n+3}}$ is divergent by the Test for Divergence.

Example 2   Determining whether the series is convergent or divergent

Determine whether the series $\sum_{n=1}^{\infty} \dfrac{\cos n\pi}{n}$ is convergent or divergent.

Solution $\{\cos n\pi\}_1^{\infty} = \{(-1)^n\}_1^{\infty}$. Thus, $\sum_{n=1}^{\infty} \dfrac{\cos n\pi}{n} = \sum_{n=1}^{\infty} (-1)^n \dfrac{1}{n}$. Let's check the two conditions for the alternating series.

1. $a_{n+1} = \dfrac{1}{n+1} \leq a_n = \dfrac{1}{n}$ for all n
2. $\lim_{n\to\infty} \dfrac{1}{n} = 0$.

Since the series satisfies the two condition, the series $\sum_{n=1}^{\infty} \dfrac{\cos n\pi}{n}$ is convergent by the Alternating Series Test.

MR. RHEE'S BRILLIANT MATH SERIES **BC ONLY** **AP CAL LESSON 40**

Example 3 Determining whether the series is convergent or divergent

Determine whether the series $\sum_{n=1}^{\infty} \dfrac{1}{e^n}$ is convergent or divergent.

Solution f is a continuous, positive, decreasing function on $[1, \infty]$. So, use the Integral Test.

$$\begin{aligned}\int_1^{\infty} e^{-x}\, dx &= \lim_{t \to \infty} \int_1^t e^{-x}\, dx \\ &= -\lim_{t \to \infty} e^{-x} \Big]_1^t \\ &= -\lim_{t \to \infty} (e^{-t} - e^{-1}) \\ &= -(e^{-\infty} - \dfrac{1}{e}) \\ &= \dfrac{1}{e}\end{aligned}$$

Since $\int_1^{\infty} e^{-x}\, dx$ is convergent, the series $\sum_{n=1}^{\infty} \dfrac{1}{e^n}$ is convergent.

Alternative solution: Consider $\sum_{n=1}^{\infty} \dfrac{1}{e^n}$ as the geometric series $\sum_{n=1}^{\infty} \left(\dfrac{1}{e}\right)^n$ with $r = \dfrac{1}{e}$. Since $|r| < 1$, the series $\sum_{n=1}^{\infty} \dfrac{1}{e^n}$ is convergent.

MR. RHEE'S BRILLIANT MATH SERIES — BC ONLY — AP CAL LESSON 40

EXERCISES

For questions 1-5, determine whether the following series are convergent or divergent.

1. $\displaystyle\sum_{n=1}^{\infty} \frac{1}{\sqrt[3]{n^2}}$

2. $\displaystyle\sum_{n=1}^{\infty} \frac{2n+1}{3n-1}$

3. $\displaystyle\sum_{n=1}^{\infty} ne^{-n^2}$

4. $\displaystyle\sum_{n=1}^{\infty} (-1)^n \frac{n}{n^2+1}$

MR. RHEE'S BRILLIANT MATH SERIES BC ONLY AP CAL LESSON 40

5. $\displaystyle\sum_{n=1}^{\infty} \frac{2^n + 3^n}{5^n}$

6. Consider the series $\displaystyle\sum_{n=1}^{\infty} n^{-2}$.

   (a) Determine whether the series converges or diverges.

   (b) Approximate the sum of the series by using the first six terms.

   (c) Find the maximum error involved in this approximation.

MR. RHEE'S BRILLIANT MATH SERIES — BC ONLY — AP CAL LESSON 40

Answers

1. Divergent
2. Divergent
3. Convergent
4. Convergent
5. Convergent
6(a). Convergent
6(b). $S_6 = 1.491$
6(c). 0.167

MR. RHEE'S BRILLIANT MATH SERIES BC ONLY AP CAL LESSON 41

LESSON 41

Convergence And Divergence Of Series, Part II

Test 6: Comparison Test

If $0 \leq a_n \leq b_n$ for n greater than some positive integer N, then the following rules apply:

1. If $\sum\limits_{n=1}^{\infty} b_n$ converges, then $\sum\limits_{n=1}^{\infty} a_n$ converges.   (known that $\sum\limits_{n=1}^{\infty} b_n$ converges in advance.)

2. If $\sum\limits_{n=1}^{\infty} a_n$ diverges, then $\sum\limits_{n=1}^{\infty} b_n$ diverges.   (known that $\sum\limits_{n=1}^{\infty} a_n$ diverges in advance.)

Tip
1. Although $\sum\limits_{n=1}^{\infty} a_n$ converges, we cannot determine whether $\sum\limits_{n=1}^{\infty} b_n$ converges or not.

2. Although $\sum\limits_{n=1}^{\infty} b_n$ diverges, we cannot determine whether $\sum\limits_{n=1}^{\infty} a_n$ diverges or not.

Test 7: Limit Comparison Test

If $\lim\limits_{n \to \infty} \dfrac{a_n}{b_n} = L$, where a_n and b_n are positive and L is a finite and positive, then either both series $\sum\limits_{n=1}^{\infty} a_n$ and $\sum\limits_{n=1}^{\infty} b_n$ converge or both diverge.

Tip You must know some series $\sum\limits_{n=1}^{\infty} b_n$ whether it converges or diverges in advance for the purpose of comparison.

MR. RHEE'S BRILLIANT MATH SERIES
BC ONLY — AP CAL LESSON 41

Test 8: Absolute Convergence Test

If a series $\sum_{n=1}^{\infty} a_n$ is absolutely convergent, then it is convergent.

Tip

1. A series $\sum_{n=1}^{\infty} a_n$ is called **absolutely convergent** if the series of absolute values $\sum_{n=1}^{\infty} |a_n|$ is convergent.

2. A series $\sum_{n=1}^{\infty} a_n$ is called **conditionally convergent** if it is convergent but not absolutely convergent.

Test 9: Ratio Test

1. If $\lim\limits_{n \to \infty} \left| \dfrac{a_{n+1}}{a_n} \right| = L < 1$, then the series $\sum_{n=1}^{\infty} a_n$ is absolutely convergent.

2. If $\lim\limits_{n \to \infty} \left| \dfrac{a_{n+1}}{a_n} \right| = L > 1$, then the series $\sum_{n=1}^{\infty} a_n$ is divergent.

3. If $\lim\limits_{n \to \infty} \left| \dfrac{a_{n+1}}{a_n} \right| = L = 1$, then the test in inconclusive. In this case, must use some other tests.

Tip

1. In general, series that involves factorials ($n!$) or nth powers (n^n) are tested using the Ratio Test.

2. For simplicity, rewrite it as shown below.

$$\lim_{n \to \infty} \left| \frac{a_{n+1}}{a_n} \right| = \lim_{n \to \infty} \left| a_{n+1} \cdot \frac{1}{a_n} \right|$$

MR. RHEE'S BRILLIANT MATH SERIES
BC ONLY — AP CAL LESSON 41

Test 10: Root Test

1. If $\lim_{n \to \infty} \sqrt[n]{|a_n|} = L < 1$, then the series $\sum_{n=1}^{\infty} a_n$ is absolutely convergent.

2. If $\lim_{n \to \infty} \sqrt[n]{|a_n|} = L > 1$, then the series $\sum_{n=1}^{\infty} a_n$ is divergent.

3. If $\lim_{n \to \infty} \sqrt[n]{|a_n|} = L = 1$, then the test in inconclusive. In this case, must use some other tests.

Example 1 Determining whether the series is convergent or divergent

Determining whether the series $\sum_{n=1}^{\infty} \dfrac{1}{\sqrt{n^3 - 1}}$ is convergent or divergent.

Solution $\sum_{n=1}^{\infty} \dfrac{1}{n^{\frac{3}{2}}}$ is a p-series with $p = \dfrac{3}{2} > 1$. So, it is convergent. Let $a_n = \dfrac{1}{\sqrt{n^3 - 1}}$ and $b_n = \dfrac{1}{n^{\frac{3}{2}}}$ and use the Limit Comparison test.

$$\lim_{n \to \infty} \frac{a_n}{b_n} = \lim_{n \to \infty} \frac{\frac{1}{\sqrt{n^3-1}}}{\frac{1}{n^{\frac{3}{2}}}} = 1 > 0$$

Since the limit is 1, the series $\sum_{n=1}^{\infty} \dfrac{1}{\sqrt{n^3 - 1}}$ is convergent.

MR. RHEE'S BRILLIANT MATH SERIES
BC ONLY AP CAL LESSON 41

Example 2 Determining whether the series is convergent or divergent

Determining whether the series $\sum_{n=1}^{\infty} \frac{\sin n}{n^2}$ is convergent or divergent.

Solution $-1 \leq \sin n \leq 1$. So, $|\sin n| \leq 1$ no matter what values of n are chosen. Thus, we use the Comparison test to compare the series $\sum_{n=1}^{\infty} \left|\frac{\sin n}{n^2}\right|$ with $\sum_{n=1}^{\infty} \frac{1}{n^2}$, which is a p-series with $p = 2 > 1$ and is convergent. Let $a_n = \left|\frac{\sin n}{n^2}\right| = \frac{|\sin n|}{n^2}$ and $b_n = \frac{1}{n^2}$. Since $a_n \leq b_n$ for all n and $\sum_{n=1}^{\infty} \frac{1}{n^2}$ is convergent, the series $\sum_{n=1}^{\infty} \frac{\sin n}{n^2}$ is absolutely convergent by the Comparison test.

Example 3 Determining whether the series is convergent or divergent

Determining whether the series $\sum_{n=1}^{\infty} \frac{(-2)^n}{n!}$ is convergent or divergent.

Solution
Since the series involves factorials ($n!$) and nth powers (2^n), let $a_n = \frac{(-1)^n 2^n}{n!}$ and $a_{n+1} = \frac{(-1)^{n+1} 2^{n+1}}{(n+1)!}$ and use the Ratio test.

$$\lim_{n \to \infty} \left|\frac{a_{n+1}}{a_n}\right| = \lim_{n \to \infty} \left|a_{n+1} \cdot \frac{1}{a_n}\right|$$
$$= \lim_{n \to \infty} \left|\frac{(-1)^{n+1} 2^{n+1}}{(n+1)!} \cdot \frac{n!}{(-1)^n 2^n}\right|$$
$$= \lim_{n \to \infty} \frac{2}{n+1}$$
$$= 0 < 1$$

Thus, the series $\sum_{n=1}^{\infty} \frac{(-2)^n}{n!}$ is absolutely convergent by the Ratio test.

MR. RHEE'S BRILLIANT MATH SERIES — BC ONLY — AP CAL LESSON 41

EXERCISES

For questions 1-6, determine whether the following series are absolutely convergent, conditionally convergent or divergent.

1. $\sum_{n=1}^{\infty} \dfrac{(-1)^n}{\sqrt{n}}$

2. $\sum_{n=1}^{\infty} \dfrac{n+1}{n^2+2}$

3. $\sum_{n=1}^{\infty} \dfrac{n!}{n^n}$

MR. RHEE'S BRILLIANT MATH SERIES   BC ONLY   AP CAL LESSON 41

4. $\sum_{n=1}^{\infty} \dfrac{1}{3^n+1}$

5. $\sum_{n=1}^{\infty} \left(\dfrac{n+1}{3n-1}\right)^n$

6. $\sum_{n=1}^{\infty} (-1)^n \dfrac{n^2 2^n}{n!}$

MR. RHEE'S BRILLIANT MATH SERIES — BC ONLY — AP CAL LESSON 41

Answers

1. Conditionally convergent
2. Divergent
3. Absolutely convergent
4. Absolutely convergent
5. Absolutely convergent
6. Absolutely convergent

LESSON 42

Strategy For Testing Series

Strategy for Testing Series

Now we have ten convergence tests for series. It is not recommended that you apply each of the ten tests to a given series until you find one works. Since there are no general rules regarding which test to apply to the given series, the following guidelines will help us save time and effort when determining the convergence of the series.

Guidelines for testing series

1. With a quick glance, if you can determine $\lim_{n\to\infty} a_n \neq 0$, then use the Test for Divergence.

2. If the series is of the form $\sum_{n=1}^{\infty} \frac{1}{n^p}$, then use the P-series test.

3. If the series is of the form $\sum_{n=1}^{\infty} ar^{n-1}$, then use the Geometric Series test.

4. If the series has a form that is similar to a p-series or a geometric series, then use the Comparison test or the Limit Comparison test.

5. If the series is of the form $\sum_{n=1}^{\infty} (-1)^n a_n$, then use the Alternating Series test.

6. If the series involves factorials ($n!$) or nth powers (n^n), then use the Ratio test.

7. If the series is of the form $\sum_{n=1}^{\infty} (a_n)^n$, then use the Root test.

8. If $a_n = f(n)$ and $\int_{1}^{\infty} f(x)\,dx$ can be easily evaluated, then use the Integral test.

MR. RHEE'S BRILLIANT MATH SERIES   BC ONLY   AP CAL LESSON 42

Example 1   Determining which test to apply to a given series

Determine which test to apply to the series $\sum_{n=1}^{\infty} ne^{-n^2}$.

Solution   Since the integral $\int_{1}^{\infty} xe^{-x^2}\, dx$ can be easily evaluated, use the Integral test.

Example 2   Determining which test to apply to a given series

Determine which test to apply to the series $\sum_{n=1}^{\infty} \frac{n^2-2}{n^2+2}$.

Solution   Since $\lim_{n\to\infty} \frac{n^2-2}{n^2+2} = 1 \neq 0$, use the Test for Divergence.

Example 3   Determining which test to apply to a given series

Determine which test to apply to the series $\sum_{n=1}^{\infty} \frac{1}{2^n+1}$.

Solution   Since the series is similar to a geometric series, use the Comparison test or the Limit Comparison test.

Example 4   Determining which test to apply to a given series

Determine which test to apply to the series $\sum_{n=1}^{\infty} (-1)^n \frac{n^2}{n^3+1}$.

Solution   The series is alternating due to $(-1)^n$, use the Alternating Series test.

MR. RHEE'S BRILLIANT MATH SERIES
BC ONLY   AP CAL LESSON 42

EXERCISES

For questions 1-10, test the series for convergence or divergence.

1. $\sum_{n=1}^{\infty} \cos\left(\frac{1}{n}\right)$

2. $\sum_{n=1}^{\infty} \frac{2^n + 1}{4^{n-1}}$

3. $\sum_{n=1}^{\infty} \frac{n+3}{\sqrt{n^3+1}}$

4. $\sum_{n=1}^{\infty} \frac{1 + \cos n}{e^n}$

5. $\sum_{n=1}^{\infty} \dfrac{2n^3 + 3n^2 + 1}{2 + 4n + 3n^3}$

6. $\sum_{n=1}^{\infty} \dfrac{n^2}{e^{n^3}}$

7. $\sum_{n=1}^{\infty} n! 2^n$

8. $\sum_{n=1}^{\infty} \dfrac{\ln n}{n}$

MR. RHEE'S BRILLIANT MATH SERIES

BC ONLY   AP CAL LESSON 42

9. $\displaystyle\sum_{n=1}^{\infty} \frac{n^2+1}{2^n+1}$

10. $\displaystyle\sum_{n=1}^{\infty} (-1)^n \frac{3^{2n}}{2^{2n}(n!)^2}$

MR. RHEE'S BRILLIANT MATH SERIES BC ONLY AP CAL LESSON 42

Answers

1. Divergent
2. Convergent
3. Divergent
4. Convergent
5. Divergent
6. Convergent
7. Divergent
8. Divergent
9. Convergent
10. Convergent

MR. RHEE'S BRILLIANT MATH SERIES   BC ONLY   AP CAL LESSON 43

LESSON 43

Power Series

Power Series

A **power series centered at a** or a **power series about a** is a series of the form

$$\sum_{n=0}^{\infty} c_n(x-a)^n = c_0 + c_1(x-a) + c_2(x-a)^2 + c_3(x-a)^3 + \cdots$$

where x is a variable and the c_n's are called **coefficients** of the series. A power series is a function of x and it may converge for some values of x and diverge for the other values of x. There are three possibilities for a given power series $\sum_{n=0}^{\infty} c_n(x-a)^n$.

1. The series converges only when $x = a$.

2. The series converges for all x.

3. There is a positive number R such that the series converges if $|x - a| < R$ and diverges if $|x - a| > R$, where R is called the **radius of convergence**. The **interval of convergence** of a power series is the interval that consists of all values of x for which the series converges.

Below summarizes the radius of convergence and interval of convergence for the three possibilities.

Possibilities	Radius of convergence	Interval of convergence
1. The series converges only when $x = a$	0	a
2. The series converges for all x	∞	$(-\infty, \infty)$
3. The series converges if $|x - a| < R$	R	$(a - R, a + R)$

When x is an endpoint of the interval, that is $x = a + R$ or $x = a - R$, the series might converge at one or both endpoints or it might diverge at both endpoints. Thus, there are four possible interval of convergence of the series:

$$(a - R, a + R) \quad [a - R, a + R) \quad (a - R, a + R] \quad [a - R, a + R]$$

Tip
1. Notice that the power series can start at $n = 0$.

2. In general, the Ratio test is used to determine the radius of convergence. Since the Ratio test fails when x is an endpoint of the interval of convergence, we must test for convergence at the endpoints with other convergence tests.

MR. RHEE'S BRILLIANT MATH SERIES
BC ONLY   AP CAL LESSON 43

Example 1   Finding the radius of convergence and interval of convergence

Finding the radius of convergence and interval of convergence of the series $\sum_{n=1}^{\infty}(-1)^n \frac{x^n}{n}$.

Solution   Let $a_n = (-1)^n \dfrac{x^n}{n}$ and use the Ratio test.

$$\begin{aligned}\lim_{n\to\infty}\left|\frac{a_{n+1}}{a_n}\right| &= \lim_{n\to\infty}\left|\frac{(-1)^{n+1}x^{n+1}}{n+1}\cdot\frac{n}{(-1)^n x^n}\right| \\ &= \lim_{n\to\infty}|x|\frac{n}{n+1} \\ &= |x|\lim_{n\to\infty}\frac{n}{n+1} \qquad\qquad \text{Since } \lim_{n\to\infty}\frac{n}{n+1}=1 \\ &= |x|\end{aligned}$$

By the Ratio test, the series converges if $|x| < 1$ and diverges if $|x| > 1$. This means that the series is centered at $a = 0$ and the radius of convergence is $R = 1$. Thus, the interval of convergence is $(-1, 1)$. we must test for convergence at the endpoints of this interval $(-1, 1)$.

When $x = -1$: $\sum_{n=1}^{\infty}(-1)^n\dfrac{(-1)^n}{n} = \sum_{n=1}^{\infty}\dfrac{1}{n}$ Diverges by the P-series test

When $x = 1$: $\sum_{n=1}^{\infty}(-1)^n\dfrac{(1)^n}{n} = \sum_{n=1}^{\infty}(-1)^n\dfrac{1}{n}$ Converges by the Alternating series test

So, the power series converges when $-1 < x \leq 1$. Therefore, the interval of convergence is $(-1, 1]$.

MR. RHEE'S BRILLIANT MATH SERIES
BC ONLY AP CAL LESSON 43

Example 2 Finding the radius of convergence and interval of convergence

Finding the radius of convergence and interval of convergence of the series $\sum_{n=1}^{\infty}(-1)^n\dfrac{(x+2)^n}{n\cdot 2^n}$.

Solution Let $a_n = (-1)^n\dfrac{(x+2)^n}{n\cdot 2^n}$ and use the Ratio test.

$$\lim_{n\to\infty}\left|\frac{a_{n+1}}{a_n}\right| = \lim_{n\to\infty}\left|\frac{(-1)^{n+1}(x+2)^{n+1}}{(n+1)\cdot 2^{n+1}}\cdot\frac{n\cdot 2^n}{(-1)^n(x+2)^n}\right|$$

$$= \frac{1}{2}|x+2|\lim_{n\to\infty}\frac{n}{n+1}$$

$$= \frac{1}{2}|x+2|$$

By the Ratio test, the series converges if $\dfrac{1}{2}|x+2| < 1$ and diverges if $\dfrac{1}{2}|x+2| > 1$. Thus, the series converges if $|x+2| < 2$ and diverges if $|x+2| > 2$. This means that the series is centered at $a = -2$ and the radius of convergence is $R = 2$. Thus, the interval of convergence is $(-4, 0)$. we must test for convergence at the endpoints of this interval $(-4, 0)$.

When $x = -4$: $\quad\sum_{n=1}^{\infty}(-1)^n\dfrac{(-4+2)^n}{n\cdot 2^n} = \sum_{n=1}^{\infty}\dfrac{1}{n}\qquad$ Diverges by the P-series test

When $x = 0$: $\quad\sum_{n=1}^{\infty}(-1)^n\dfrac{(0+2)^n}{n\cdot 2^n} = \sum_{n=1}^{\infty}(-1)^n\dfrac{1}{n}\qquad$ Converges by the Alternating series test

So, the power series converges when $-4 < x \leq 0$. Therefore, the interval of convergence is $(-4, 0]$.

MR. RHEE'S BRILLIANT MATH SERIES   BC ONLY   AP CAL LESSON 43

EXERCISES

For questions 1-6, find the radius of convergence and interval of convergence of the following series.

1. $\sum_{n=1}^{\infty} \dfrac{n(x-2)^n}{5^{n-1}}$

2. $\sum_{n=1}^{\infty} (-1)^n \dfrac{n(x+3)^n}{4^n}$

3. $\sum_{n=1}^{\infty} (-1)^n \dfrac{x^n}{2n+1}$

MR. RHEE'S BRILLIANT MATH SERIES   BC ONLY   AP CAL LESSON 43

4. $\displaystyle\sum_{n=1}^{\infty} \frac{(-3)^n x^n}{\sqrt{n+1}}$

5. $\displaystyle\sum_{n=1}^{\infty} (-1)^n \frac{x^{2n}}{(2n)!}$

6. $\displaystyle\sum_{n=1}^{\infty} \frac{(3x+2)^n}{n^2}$

MR. RHEE'S BRILLIANT MATH SERIES
BC ONLY — AP CAL LESSON 43

Answers

1. ROC = 5, IOC = $(-3, 7)$
2. ROC = 4, IOC = $(-7, 1)$
3. ROC = 1, IOC = $(-1, 1]$
4. ROC = $\frac{1}{3}$, IOC = $\left(-\frac{1}{3}, \frac{1}{3}\right]$
5. ROC = ∞, IOC = $(-\infty, \infty)$
6. ROC = $\frac{1}{3}$, IOC = $\left[-1, -\frac{1}{3}\right]$

MR. RHEE'S BRILLIANT MATH SERIES BC ONLY AP CAL LESSON 44

LESSON 44

Representations Of Functions As Power Series

Representations of Functions as Power Series

We will learn how to represent functions as power series by using a geometric series or by differentiating or integrating a series. This will be very useful for integrating functions that do not have antiderivatives, and for approximation.

Recall that given the geometric series $\sum_{n=1}^{\infty} ar^{n-1} = a + ar + ar^2 + \cdots$, it is convergent if $|r| < 1$, and its sum is $\dfrac{a}{1-r}$. Otherwise, the geometric series is divergent if $|r| \geq 1$. So, If you start with a geometric series $\sum_{n=0}^{\infty} x^n$ whose first term is 1 and the common ratio $r = x$

$$\sum_{n=0}^{\infty} x^n = 1 + x + x^2 + x^3 + \cdots = \frac{1}{1-x}, \quad \text{if } |x| < 1$$

the sum of the geometric series is $\dfrac{1}{1-x}$. Thus, the representation of the function $\dfrac{1}{1-x}$ with power series can be given by $\dfrac{1}{1-x} = \sum_{n=0}^{\infty} x^n$

Some algebraic manipulations are required to represent functions as power series. Below shows how to do algebraic manipulations to find the power series representations for some functions.

$$\frac{1}{1-x} = \sum_{n=0}^{\infty} x^n$$

$$\frac{1}{1-x^2} = \sum_{n=0}^{\infty} (x^2)^n = \sum_{n=0}^{\infty} x^{2n}$$

$$\frac{1}{1+x} = \frac{1}{1-(-x)} = \sum_{n=0}^{\infty} (-x)^n = \sum_{n=0}^{\infty} (-1)^n x^n$$

$$\frac{1}{1+x^2} = \frac{1}{1-(-x^2)} = \sum_{n=0}^{\infty} (-x^2)^n = \sum_{n=0}^{\infty} (-1)^n x^{2n}$$

$$\frac{1}{2+x} = \frac{1}{2\left(1+\frac{x}{2}\right)} = \frac{1}{2\left(1-\left(-\frac{x}{2}\right)\right)} = \frac{1}{2}\sum_{n=0}^{\infty} \left(-\frac{x}{2}\right)^n = \sum_{n=0}^{\infty} \frac{(-1)^n}{2^{n+1}} x^n$$

$$\frac{x^2}{1+x} = x^2 \cdot \frac{1}{1+x} = x^2 \cdot \frac{1}{1-(-x)} = x^2 \cdot \sum_{n=0}^{\infty} (-x)^n = \sum_{n=0}^{\infty} (-1)^n x^{n+2}$$

MR. RHEE'S BRILLIANT MATH SERIES — BC ONLY — AP CAL LESSON 44

Differentiation and Integration of Power Series

If the power series $\sum_{n=0}^{\infty} c_n(x-a)^n$ has radius of convergence $R > 0$, then the function f defined by

$$f(x) = c_0 + c_1(x-a) + c_2(x-a)^2 + c_3(x-a)^3 + \cdots = \sum_{n=0}^{\infty} c_n(x-a)^n$$

is differentiable on the interval $(a - R, a + R)$. Thus,

$$f'(x) = c_1 + 2c_2(x-a) + 3c_3(x-a)^2 + \cdots = \sum_{n=1}^{\infty} nc_n(x-a)^{n-1}$$

and

$$\int f(x)\, dx = C + c_0(x-a) + c_1\frac{(x-a)^2}{2} + c_2\frac{(x-a)^3}{3} + \cdots = C + \sum_{n=0}^{\infty} c_n \frac{(x-a)^{n+1}}{n+1}$$

Below shows how to represent functions as power series using differentiation and integration.

$$\frac{1}{(1-x)^2} = \left(\frac{1}{1-x}\right)' = \left(\sum_{n=0}^{\infty} x^n\right)' = \sum_{n=1}^{\infty} nx^{n-1}$$

$$\ln(1-x) = -\int \frac{1}{1-x}\, dx = -\int \sum_{n=0}^{\infty} x^n\, dx = C - \sum_{n=0}^{\infty} \frac{x^{n+1}}{n+1}$$

$$\tan^{-1} x = \int \frac{1}{1+x^2}\, dx = \int \sum_{n=0}^{\infty} (-1)^n x^{2n}\, dx = C + \sum_{n=0}^{\infty} (-1)^n \frac{x^{2n+1}}{2n+1}$$

Tip: In order to differentiate or integrate a power series, differentiate or integrate each individual term in the power series. This is called **term-by-term differentiation and integration**.

MR. RHEE'S BRILLIANT MATH SERIES — BC ONLY — AP CAL LESSON 44

Example 1 Finding a power series representation for the function

Find a power series representation for the function $f(x) = \dfrac{x}{4-x}$.

Solution

$$\begin{aligned}
\frac{x}{4-x} &= x \cdot \frac{1}{4-x} \\
&= x \cdot \frac{1}{4\left(1 - \dfrac{x}{4}\right)} \\
&= x \cdot \frac{1}{4} \sum_{n=0}^{\infty} \left(\frac{x}{4}\right)^n \\
&= \sum_{n=0}^{\infty} \frac{1}{4^{n+1}} x^{n+1}
\end{aligned}$$

Example 2 Finding a power series representation for the function

Find a power series representation for the function $f(x) = x\ln(1+x)$.

Solution

$$\begin{aligned}
x\ln(1+x) &= x \cdot \ln(1+x) = x \cdot \int \frac{1}{1+x}\, dx \\
&= x \cdot \int \frac{1}{1-(-x)}\, dx = x \cdot \int \sum_{n=0}^{\infty} (-x)^n\, dx \\
&= x \cdot \int \sum_{n=0}^{\infty} (-1)^n x^n\, dx \\
&= x \cdot \sum_{n=0}^{\infty} (-1)^n \frac{x^{n+1}}{n+1} + C \\
&= C + \sum_{n=0}^{\infty} (-1)^n \frac{x^{n+2}}{n+1}
\end{aligned}$$

MR. RHEE'S BRILLIANT MATH SERIES
BC ONLY   AP CAL LESSON 44

EXERCISES

For questions 1-6, find the power series representations for the following functions.

1. $f(x) = \dfrac{1+x}{1-x}$

2. $f(x) = \dfrac{1}{1+2x}$

3. $f(x) = \dfrac{1}{(1-x)^3}$

MR. RHEE'S BRILLIANT MATH SERIES   BC ONLY   AP CAL LESSON 44

4. $f(x) = \dfrac{1}{(1+2x)^2}$

5. $f(x) = \tan^{-1} 2x$

6. $f(x) = \ln(4-x)$

MR. RHEE'S BRILLIANT MATH SERIES BC ONLY AP CAL LESSON 44

Answers

1. $-1 + 2\sum_{n=0}^{\infty} x^n$

2. $\sum_{n=0}^{\infty} (-1)^n 2^n x^n$

3. $\dfrac{1}{2}\sum_{n=2}^{\infty} n(n-1)x^{n-2}$

4. $\sum_{n=2}^{\infty} (-1)^{n-1} n 2^{n-1} x^{n-1}$

5. $C + \sum_{n=0}^{\infty} (-1)^n \dfrac{2^{2n+1} x^{2n+1}}{2n+1}$

6. $C - \sum_{n=0}^{\infty} \dfrac{x^{n+1}}{4^{n+1}(n+1)}$

LESSON 45
Taylor And Maclaurin Series

Taylor Series of the function f at a

Suppose the function f has a power series representation at a. Then, f is given by

$$f(x) = \sum_{n=0}^{\infty} \frac{f^{(n)}(a)}{n!}(x-a)^n = f(a) + \frac{f'(a)}{1!}(x-a) + \frac{f''(a)}{2!}(x-a)^2 + \frac{f'''(a)}{3!}(x-a)^3 + \cdots$$

where $f^{(n)}(a)$ is the nth-order derivative at a and the series is called the **Taylor series of the function f at a** (or **about a** or **centered at a**).

Maclaurin Series

If we use $a = 0$, then the Taylor series at $a = 0$ becomes

$$f(x) = \sum_{n=0}^{\infty} \frac{f^{(n)}(0)}{n!} x^n = f(0) + \frac{f'(0)}{1!}x + \frac{f''(0)}{2!}x^2 + \frac{f'''(0)}{3!}x^3 + \cdots$$

and the series is called the **Maclaurin series**.

Below shows some important Maclaurin series that you need to know.

$$\frac{1}{1-x} = \sum_{n=0}^{\infty} x^n = 1 + x + x^2 + x^3 + \cdots$$

$$e^x = \sum_{n=0}^{\infty} \frac{x^n}{n!} = 1 + \frac{x}{1!} + \frac{x^2}{2!} + \frac{x^3}{3!} + \cdots$$

$$\sin x = \sum_{n=0}^{\infty} (-1)^n \frac{x^{2n+1}}{(2n+1)!} = x - \frac{x^3}{3!} + \frac{x^5}{5!} - \frac{x^7}{7!} + \cdots$$

$$\cos x = \sum_{n=0}^{\infty} (-1)^n \frac{x^{2n}}{(2n)!} = 1 - \frac{x^2}{2!} + \frac{x^4}{4!} - \frac{x^6}{6!} + \cdots$$

$$\tan^{-1} x = \sum_{n=0}^{\infty} (-1)^n \frac{x^{2n+1}}{(2n+1)} = x - \frac{x^3}{3} + \frac{x^5}{5} - \frac{x^7}{7} + \cdots$$

MR. RHEE'S BRILLIANT MATH SERIES BC ONLY AP CAL LESSON 45

Example 1 Finding the Maclaurin series of the function

(a) Find the Maclaurin series for the function $f(x) = e^x$.

(b) Find the Maclaurin series for the function $f(x) = e^{-2x}$ using the series obtained in part (a).

Solution

(a) If $f(x) = e^x$, its nth-order derivatives are e^x. Thus, $f^{(n)}(0) = e^0 = 1$.

$$e^x = \sum_{n=0}^{\infty} \frac{f^{(n)}(0)}{n!}(x)^n = \sum_{n=0}^{\infty} \frac{x^n}{n!}$$

(b) In order to find Maclaurin series for the function $f(x) = e^{-2x}$, substitute $-2x$ into x in $e^x = \sum_{n=0}^{\infty} \frac{x^n}{n!}$. Thus,

$$e^{-2x} = \sum_{n=0}^{\infty} \frac{(-2x)^n}{n!} = \sum_{n=0}^{\infty} (-1)^n \frac{2^n x^n}{n!}$$

Example 2 Fining the Taylor series of the function.

Find the Taylor series of $f(x) = \ln x$ centered at $a = 2$.

MR. RHEE'S BRILLIANT MATH SERIES
BC ONLY — AP CAL LESSON 45

Solution We need to calculate $\dfrac{f^{(n)}(2)}{n!}$ first.

When $n = 0$: $\quad f(x) = \ln x \quad\quad f(2) = \ln 2 \quad\quad \dfrac{f(2)}{0!} = \ln 2$

When $n = 1$: $\quad f'(x) = \dfrac{1}{x} \quad\quad f'(2) = \dfrac{1}{2} \quad\quad \dfrac{f'(2)}{1!} = \dfrac{1}{2 \cdot 1!}$

When $n = 2$: $\quad f''(x) = -\dfrac{1}{x^2} \quad\quad f''(2) = -\dfrac{1}{2^2} \quad\quad \dfrac{f''(2)}{2!} = -\dfrac{1}{2^2 \cdot 2!}$

When $n = 3$: $\quad f'''(x) = \dfrac{2!}{x^3} \quad\quad f'''(2) = \dfrac{2!}{2^3} \quad\quad \dfrac{f'''(2)}{3!} = \dfrac{2!}{2^3 \cdot 3!}$

Thus, the Taylor series of $f(x) = \ln x$ centered at $a = 2$ is

$$\ln x = \sum_{n=0}^{\infty} \frac{f^{(n)}(2)}{n!} (x-2)^n$$

$$= \ln 2 + \frac{1}{2}(x-2) - \frac{1}{8}(x-2)^2 + \frac{1}{24}(x-2)^3 + \cdots$$

$$= \ln 2 + \sum_{n=1}^{\infty} (-1)^{n+1} \frac{1}{2^n \cdot n} (x-2)^n$$

Approximating the function f using Taylor polynomial

The Taylor series of the function f at a, $\displaystyle\sum_{n=0}^{\infty} \frac{f^{(n)}(a)}{n!} (x-a)^n$, is an infinite series. In order to approximate the function of f, it's not practical to include infinite terms of the Taylor series. Instead, we use the **nth-degree Taylor polynomial of f, T_n**, which is given by

$$f(x) \approx T_n(x) = \sum_{k=0}^{n} \frac{f^{(k)}(a)}{k!}(x-a)^k = f(a) + \frac{f'(a)}{1!}(x-a) + \frac{f''(a)}{2!}(x-a)^2 + \cdots + \frac{f^{(n)}(a)}{n!}(x-a)^n$$

$R_n(x)$ is the called the **remainder** (error) involved in approximating the function $f(x)$ by the nth-degree Taylor polynomial, $T_n(x)$ and is defined by

$$R_n(x) = f(x) - T_n(x)$$

As $n \to \infty$, $T_n(x) \approx f(x)$. Thus, $R_n \to 0$.

MR. RHEE'S BRILLIANT MATH SERIES   BC ONLY   AP CAL LESSON 45

Lagrange Error Bound

If T_n is the nth-degree Taylor polynomial for $f(x)$ centered at a, then the remainder (error) $R_n = f(x) - T_n(x)$ is bounded by

$$|R_n(x)| \leq \frac{M}{(n+1)!}|x-a|^{n+1}$$

where M is some value satisfying $|f^{(n+1)}(x)| \leq M$ on the interval between a and x.

1. $|R_n(x)| \leq \dfrac{f^{(n+1)}(x)}{(n+1)!}|x-a|^{n+1}$ means that the upper bound for the error is the absolute value of the first term of the omitted terms not being used for approximation.

2. The Lagrange error bound gives us an upper bound for the error. It does not give exact value of the error.

Example 3   Finding the Taylor polynomial of the function and error

(a) Find the second-degree Taylor polynomial of $\ln(1+x)$ at $a = 0$.

(b) Use this to estimate $\ln(1.2)$.

(c) Estimate the error in this approximation.

MR. RHEE'S BRILLIANT MATH SERIES
BC ONLY   AP CAL LESSON 45

Solution

(a) Finding the Maclaurin series directly from $\sum_{n=0}^{\infty} \dfrac{f^{(n)}(0)}{n!} x^n$ is time consuming. Instead,

$$\ln(1+x) = \int \frac{1}{1+x} \, dx = \int \frac{1}{1-(-x)} \, dx$$

$$= \int \sum_{n=0}^{\infty} (-1)^n x^n \, dx$$

$$= C + \sum_{n=0}^{\infty} (-1)^n \frac{x^{n+1}}{n+1}$$

Substituting $x = 0$ into $\ln(1+x) = C + \sum_{n=0}^{\infty} (-1)^n \dfrac{x^{n+1}}{n+1}$ gives $C = 0$. Thus,

$$\ln(1+x) = \sum_{n=0}^{\infty} (-1)^n \frac{x^{n+1}}{n+1} = x - \frac{x^2}{2} + \frac{x^3}{3} - \frac{x^4}{4} + \cdots$$

Therefore, the second-degree Taylor polynomial, $T_2(x)$, is

$$\ln(1+x) \approx T_2(x) = x - \frac{x^2}{2}$$

(b) If $f(x) = \ln(1+x)$, $f(0.2) = \ln(1.2)$. Thus,

$$f(0.2) = \ln(1.2) \approx T_2(0.2) = (0.2) - \frac{(0.2)^2}{2} = 0.18$$

(c) $R_2(x) = f(x) - T_2(x)$. According to the Lagrange error bound,

$$|R_2(x)| \le \frac{M}{3!} |x|^3, \qquad \text{where } |f^{(3)}(x)| \le M$$

$$|R_2(0.2)| \le \frac{M}{3!} |0.2|^3$$

In order to find M satisfying $|f^{(3)}(x)| \le M$, we need to find $f^{(3)}(x)$.

$$f'(x) = \frac{1}{1+x}, \qquad f''(x) = -\frac{1}{(1+x)^2}, \qquad f^{(3)}(x) = \frac{2}{(1+x)^3}$$

300

Since $f^{(3)}(x) = \dfrac{2}{(1+x)^3}$ is decreasing on the interval $[0, 0.2]$, the value of $f^{(3)}(x)$ is greatest at $x = 0$ on the interval. So, $M = f^{(3)}(0) = 2$. Thus,

$$\left| R_2(0.2) \right| \leq \frac{M}{3!} |0.2|^3 = \frac{2(0.2)^3}{3!} = 0.00267$$

Therefore, the error in this approximation is 0.00267.

MR. RHEE'S BRILLIANT MATH SERIES — BC ONLY — AP CAL LESSON 45

EXERCISES

1. Find the Maclaurin series for $f(x) = \dfrac{\sin x}{x}$.

2. Find the Maclaurin series for $f(x) = x^2 e^{-x}$.

3. Find the Maclaurin series for $f(x) = x \sin 2x$.

MR. RHEE'S BRILLIANT MATH SERIES   BC ONLY   AP CAL LESSON 45

4. Find the Taylor series for $f(x) = e^x$ at $a = 3$.

5. Find the first four terms and the general terms of the Taylor series for $f(x) = \ln x$ centered at $a = 1$.

6. The first three derivatives of $f(x) = (x+1)^{\frac{5}{2}}$ are
$$f'(x) = \frac{5}{2}(x+1)^{\frac{3}{2}}, \qquad f''(x) = \frac{15}{4}(x+1)^{\frac{1}{2}}, \qquad f^{(3)}(x) = \frac{15}{8(x+1)^{\frac{1}{2}}}$$

(a) Find the second-degree Taylor polynomial at $a = 0$ for $f(x)$.

(b) Use the polynomial in part (a) to approximate $f\left(\dfrac{1}{2}\right)$.

(c) Estimate the error in this approximation.

MR. RHEE'S BRILLIANT MATH SERIES — BC ONLY — AP CAL LESSON 45

Answers

1. $\displaystyle\sum_{n=0}^{\infty}(-1)^n \frac{x^{2n}}{(2n+1)!}$

2. $\displaystyle\sum_{n=0}^{\infty}(-1)^n x^{n+2}$

3. $\displaystyle\sum_{n=0}^{\infty}(-1)^n \frac{2^{2n+1} x^{2n+2}}{(2n+1)!}$

4. $\displaystyle\sum_{n=0}^{\infty} \frac{e^3}{n!}(x-3)^n$

5. $T_4(x) = \displaystyle\sum_{n=1}^{4}(-1)^{n+1}\frac{(x-1)^n}{n} = (x-1) - \frac{(x-1)^2}{2} + \frac{(x-1)^3}{3} - \frac{(x-1)^4}{4}$

6(a). $T_2(x) = 1 + \dfrac{5}{2}x + \dfrac{15}{8}x^2$

6(b). 2.719

6(c). 0.039